Blockchain for Cyberphysical Systems

For a complete listing of titles in the
Artech House Information Security and Privacy Seriess,
turn to the back of this book.

Blockchain for Cyberphysical Systems

Ali Dorri

Salil Kanhere

Raja Jurdak

ARTECH HOUSE

BOSTON | LONDON

artechhouse.com

Library of Congress Cataloging-in-Publication Data
A catalog record for this book is available from the U.S. Library of Congress.

British Library Cataloguing in Publication Data
A catalogue record for this book is available from the British Library.

Cover design by John Gomes

ISBN 13: 978-1-63081-783-1

685 Canton Street
Norwood, MA 02062

10 9 8 7 6 5 4 3 2 1

For Aude, Sophie, Marc, Muna, Murad, and Hania
—Raja Jurdak

For Alpa, Aai, Baba, Mummy, Papa, Nidhi, Rushabh, Param, and Yuveer
—Salil Kanhere

For Parisa, Mom, Dad, Fatemeh, and Mohammad Hossein
—Ali Dorri

Contents

Preface

This book covers a journey that we started in 2016 to investigate the potential of blockchain for cyberphysical systems (CPS). At the time, Dr. Kanhere and Dr. Jurdak had been involved in Internet of Things (IoT) research over many years, and blockchain had been around for several years as the underpinning technology for Bitcoin and other cryptocurrencies. However, within a cyberphysical context, blockchain's potential was underexplored. In March 2016, Dr. Dorri started his Ph.D. at UNSW under the supervision of Dr. Kanhere and Dr. Jurdak, and the focus of his Ph.D. research was blockchain in the IoT. This research quickly led to a position paper in 2016, followed by a series of foundational papers that proposed a lightweight, scalable blockchain for IoT. We also investigated related topics, such as how to remove blocks from immutable blockchains to save memory and preserve privacy in large scale CPS, and the application of blockchain in cyberphysical scenarios ranging from connected and autonomous vehicles to smart grids.

Along this journey, several additional Ph.D. students and researchers joined the team across Brisbane and Sydney to investigate new directions of blockchain for CPS, investigating horizontal challenges such as anonymity in blockchain transactions that store sensor data to trust in cyberphysical data and entities, and new applications from traceability in sensor-supported supply chains, to IoT data marketplaces. These new directions enriched our research in this area and helped to identify new challenges and opportunities for blockchain in CPS. At the time of this writing, blockchain for CPS is a flourishing research area with a growing researcher base globally. We hope that we have played a part in the growth and uptake of this area.

This book is the culmination of our 4-year journey in developing blockchain for CPS with the aim to share with readers the steps in our journey and the learnings we derived from it. The book is structured into 3 parts. Part I introduces CPS, their challenges, opportunities, and current solutions and the potential of blockchain in this context. Part II discusses the horizontal blockchain challenges and solutions for CPS, such as scalability, immutability, anonymity, and trust, and Part III covers vertical applications of blockchain for CPS. We have very much enjoyed our journey so far in this area. We hope the readers will likewise enjoy this book.

Acknowledgments

We would like to thank Volkan Dedeoglu, Sidra Malik, Chuka Oham, Fengji Luo, Gianluca Ranzi, and Pooja Gupta for their authorship and contribution for some chapters of this book, as well as for their contributions to our team's research over the past several years. We would also like to thank our collaborators on some of the works that have underpinned this book, including Sanjay Jha, ZhaoYang Dong, Praveen Gauravaram, Clemence Roulin, Regio Michelin, Marco Steger, Kamran Najeebullah, Roben Castagna Lunardi, and Avelino Zorzo.

Part I

1

Introduction to Cyberphysical Systems

1.1 Introduction

Cyberphysical systems (CPS) have now become an integral part of daily life, thanks to advances in miniaturized sensors, low-power processors, and radios. Examples of CPS include security and fire alarm systems in our homes, automated supply chains of the food that we consume, and the increasingly autonomous and connected cars that we drive. The growing appetite for CPS adoption has been driven by an increased role of digital technologies in our lives and by the need to interact with and automate these technologies more effectively.

So, what exactly are CPS? CPS integrate sensors, processors, and actuators in order to acquire data from our physical surroundings, analyze it, and then make some decisions or take some actions. CPS are often described as systems that sense, think, and act across the physical and digital domains, as shown in Figure 1.1. A critical feature of CPS is the closed loop shown in Figure 1.1, where decisions and actions made previously can affect the physical context, which can determine the next set of sensed observations.

CPS typically involve one or more sensors to capture observations from the physical world. An example of CPS consisting of two nodes is shown in Figure 1.2, where node A has one sensor, and node B has two sensors to observe the physical context, as well as an actuator to take some action.

The sensor observation data of the physical context is transformed into actionable information through algorithms that run on one or more processors in order to continuously assess the state of the physical world. The results are then used to guide decisions and actions. In the example of Figure 1.2, nodes

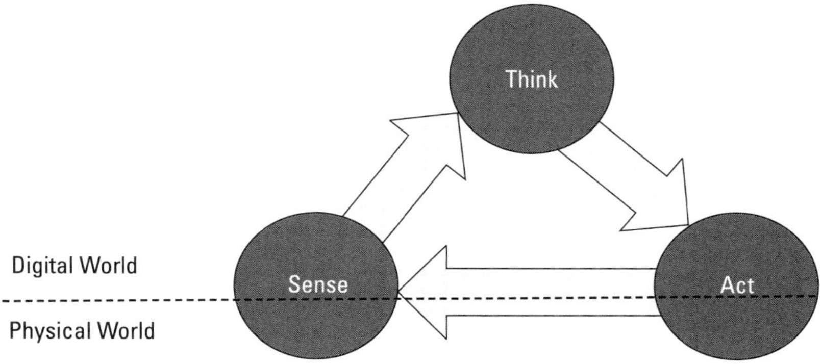

Figure 1.1 CPS sense, think, and act.

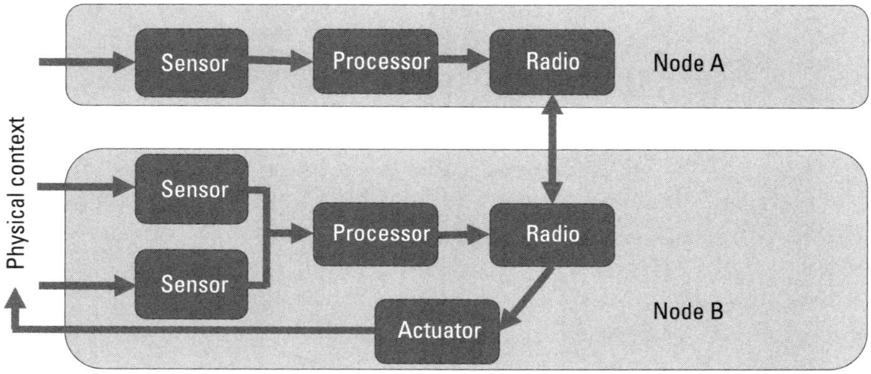

Figure 1.2 Information flow in the CPS sense-think-act cycle.

A and B need to communicate using their radios to share the outputs of their algorithms in order to reach a decision based on the shared observations across the two nodes. CPS nodes can then decide to take actions. These actions may involve a change in the configuration of one or more nodes in the CPS, such as updating the system parameters, or actuation back in the physical domain, such as the movement of a robotic component within the system. In the latter case, CPS actions can alter the state of the physical context, which then impacts the next set of observations for the CPS nodes. Through their sensing and actuation, CPS are tightly coupled to their physical contexts, which often require real-time processing and in situ decisions.

The above discussion has presented the generic information flow in CPS. Let's now illustrate a more applied CPS example involving distributed control of heating, ventilation, and air conditioning (HVAC) systems [1], illustrated in Figure 1.3. The goal of this system is to ensure the thermal comfort of a

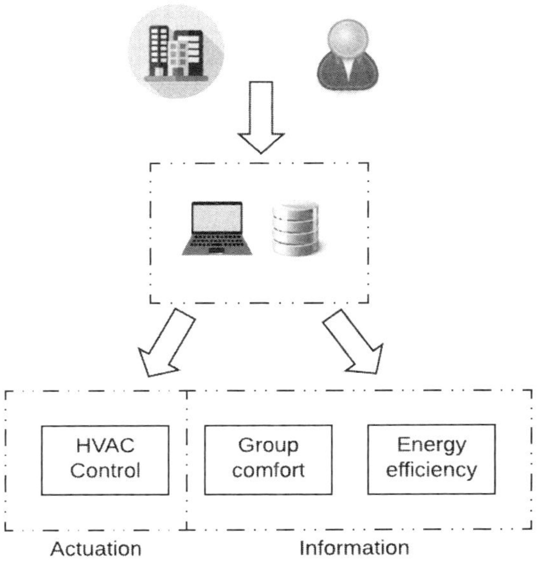

Figure 1.3 A distributed user-driven HVAC control system.

building's current occupants, while also ensuring that the system's operation is energy-efficient. The idea is to acquire frequent data from various sensors to determine current building occupants and their thermal comfort zones in order to control the HVAC system dynamically to achieve both the comfort and energy efficiency goals. For instance, if occupants are flexible in terms of their thermal comfort, the system can reduce the cooling intensity of HVAC to save energy.

A particular instance of this CPS integrates data from the building's management system, as well as occupancy and thermal preference data through an application installed on users' smartphones [2]. Users can periodically vote through their phones on whether they are too cold or too hot to feed into the system's decisions, while the system can automatically detect the presence of users in the building through the phone's wireless interface. Once occupancy and thermal comfort preferences of occupants are determined, the system determines the best HVAC set point temperature that satisfies the comfort and energy efficiency requirements. The new HVAC set point then changes the building temperature. Over the next voting period, the building occupancy may change and the current occupants can then vote again through the app on their thermal comfort for this period, which will determine the system's HVAC set point for the next round. This system demonstrates the tight feedback loop between the physical and digital world in CPS: a group of smartphones, acting as virtual comfort and occupancy sensors, can affect the decisions of the CPS.

The CPS's decisions then determine the observations of the sensors in the next round, which drive the CPS's next decisions, and so on. The value of CPS here is to enable a large group of users to dynamically input their thermal comfort preferences in a distributed manner, which feed into the decisions of CPS.

It is worth noting that CPS are closely related to the Internet of Things (IoT), which is a network of connected devices with sensing, computing, communication, and actuation capabilities. While both CPS and IoT emphasize node connectivity and interaction, CPS is particularly concerned with the interface between digital and physical realms for sensing and actuation and the focus is on addressing the challenges that arise at this interface. In this book, we use the term CPS for broad representation of scenarios at this digital/physical interface, and we refer to IoT as the interconnecting network and distributed interactions that enable CPS entities to communicate and coordinate at scale.

In the rest of this chapter, we first introduce some of the applications of CPS in Section 1.2. Section 1.3 outlines the key challenges involved in CPS. Finally, Section 1.4 concludes the chapter.

1.2 CPS Application Opportunities

In this section, we outline some representative application opportunities of CPS as a means to highlight the common design choices and challenges for CPS.

The potential of CPS is immense across many economic sectors. For this reason, CPS underpins the fourth industrial revolution [3] that promises transformation of industries and consumer markets through sensing, robotics, artificial intelligence, and other emerging technologies. Thanks to integrated sensors, CPS are context-aware as they can sense changes in physical context and adapt system behavior accordingly. CPS are also closed-loop, and sensing, processing, and decision-making are autonomous, which enables in situ unattended operation and greater operational efficiency and responsiveness to dynamic situations. These benefits can be used for real-time decisions, such as fault detection in various applications. CPS benefits have materialized in many sectors including agriculture [4], health [5], manufacturing [6], supply chains [7], ecology [8], and energy [9].

CPS cover a broad range of technologies and instantiations, ranging from standalone CPS such as a robotic platform to networked CPS that can involve a large number of resource-constrained devices. A diverse range of applications, heterogeneity of the devices, and ability to exchange data with any device are among the key factors that make the development of CPS challenging. Table 1.1 provides an overview of key application areas of CPS and maps their relevant challenges.

Table 1.1
Mapping CPS Sectors Against Key Challenges: Most the Challenges Are Common to All Sectors

	Centralization	Security	Privacy	Heterogeneity	Auditability	Persistence	Trust
Agriculture	√	√	√	√	√	√	√
Health	√	√	√	√	√	—	√
Manufacturing	√	√	√	√	√	√	√
Supply Chain	√	√	√	√	√	√	√
Ecology	√	√	—	√	—	√	√
Defense	√	√	√	√	√	√	√
Energy	√	√	√	√	√	—	√
Transport	√	√	√	√	√	—	√

In the remainder of this section, we discuss applications of CPS in smart cities (Section 1.2.1), smart grids (Section 1.2.2), and supply chains (Section 1.2.3) as representative CPS applications. For a comprehensive discussion of CPS applications, we refer readers to [10, 11].

1.2.1 Smart Cities

A smart city encompasses a broad range of sensors that sense urban areas and use the data provided by the sensors along with the data of participants (e.g., citizens) to offer efficient personalized services to the citizens and manage the city resources. For instance, smart connected vehicles are equipped with a broad range of sensors, including Global Positioning System (GPS), dashboard cameras, and Light Detection and Ranging (LIDAR), that produce large volume of data, which is estimated to reach 4,000 GB per day [12]. The huge volume of data produced by CPS devices is processed in the service providers (SPs) to offer personalized services to the users (e.g., online mapping or finding the best parking spots).

High connectivity of the participating nodes in a smart city introduces a broad range of services, but on the flip side, also raises security risks as attackers may compromise devices, which may lead to safety threats. As connectivity becomes more pervasive in vehicular systems and more entities (e.g., service centers) are involved during the operational life cycle of a vehicle, addressing the liability challenge in case of an accident is challenging. The smart vehicles collect privacy-sensitive data about the vehicle owner (e.g., the location of the vehicle) and exchange with SPs. This potentially enables the SPs to build virtual profiles about the vehicle owners and, in turn, compromise the privacy of the users. We will further discuss the challenges in smart cities in Chapter 9.

1.2.2 Smart Grids

A smart grid consists of nodes that collaborate to respond to the energy demand of the grid users. These nodes can be energy consumers, producers, transmission infrastructure, smart meters, and renewable sources of energy. In recent years, distributed sources of energy (e.g., solar panels) penetrated the smart grid which revolutionized the way energy is generated and fed in the smart grid. The emergence of the two-way communication model, where the participating nodes can receive or inject energy to the grid, coupled with CPS technology accelerates the growth of smart grids. The CPS devices enable real-time monitoring of the energy demand and generation, predicting the future demand of the grid, trading the energy between multiple participants, and balancing the load using signals from the energy companies. As an example, an energy company may decide to increase the energy price during the peak demand period in real time. Smart appliances in a home may consequently adjust their operation to reduce the energy cost for the end user.

The existing smart grid infrastructure relies on centralized brokered communications that suffers from single point of failure and many-to-one traffic nature. The penetration of distributed energy sources along with the introduction of a broad range of CPS devices introduces new communication models, which increase the packet and bandwidth overhead in conventional methods and limit their scalability. The broker has the full control to identify the energy price even if end-users agree to trade energy between each other; for example, a smart home user that has solar panels on his roof may trade the excess energy with their neighbor. The central authorities can employ the collected information about the users to build virtual profiles of them, which may reveal private information of the users (e.g., the energy consumption or generation pattern) and thus compromises their privacy. We will further study the smart grid and the corresponding challenges in Chapter 8.

1.2.3 Supply Chains

Supply chains maintain the information about a product (or service) as it makes it way from the supplier to the customer. CPS devices are often installed in various stages in the supply chain to monitor the environment; for example, temperature sensors are employed in a food supply chain to ensure that food products are kept at the proper temperature as they make their way through the various entities from the producer to the consumer. CPS sensors are aimed at automation of data collection from supply chains as a step towards transparency and auditability. As traded goods and services in supply chains are highly diverse, the sensing devices used to track their state cover a broad range from temperature, humidity, and pressure to other, more product-specific indicators. Supply chains are inherently decentralized and include multiple trading entities that are both cooperative and competitive in some instances, which highlights the need for trust among the participants. When CPS are used for automated data collection in supply chains, the issue of data security and trust that the data is authentic and truly representative of the physical state of traded goods are open challenges.

1.3 CPS Challenges

Realizing the many opportunities offered by CPS involves unique challenges due to the tight coupling with the physical context and environment. In this section, we discuss the challenges in CPS, which include centralization (see Section 1.3.1), security (see Section 1.3.2), privacy (see Section 1.3.3), heterogeneity in devices (see Section 1.3.4), lack of control or auditability over or data (see Section 1.3.5), persistence/sustainability (see Section 1.3.6), and trust (see Section 1.3.7).

1.3.1 Centralization

The existing communication architectures for CPS, including the IoT, rely on centralized brokered communications where the devices can only communicate through central controllers (e.g., the cloud servers of an SP). These architectures employ the client-server communication model and the central server is ultimately responsible for performing many of the coordination tasks including identifying, connecting, and authorizing the devices. This potentially limits the scalability of CPS systems where bandwidth and processing requirements at the central server increase with the number of nodes in a CPS system. This can lead to delays and service level deterioration for the participating nodes. The client-server communication model leads to many-to-one traffic patterns where the network traffic is sent to a single entity and creates a single point of failure challenge. According to the latter, if the server fails, the whole network becomes inaccessible. In current centralized architectures, two devices located nearby often have to communicate through their respective central servers, which is highly inefficient and leads to bandwidth and delay overheads.

The potential value of data collected by CPS, as independently deployed systems, can lead to the creation of data silos. In this scenario, two or more independent CPS systems whose data could have a shared useful purpose deposit this data at their own central server, which quarantines this data from sharing and misses the opportunities for value creation through data sharing.

1.3.2 Security

As CPS systems are maturing, the key focus of research and development has been in developing their key functionalities, such as autonomous operation and reliable communication. The security of CPS has so far been a secondary consideration, in many cases, due to the limited computational capacity of CPS nodes to support additional security mechanisms beyond their core application function. Additionally, CPS encompasses heterogeneous devices from different manufacturers with poor or no built-in security features. This has created vulnerabilities in CPS instantiations at different scales, ranging from tiny IoT devices to connected vehicles. For instance, the Mirai botnet attack compromised around 145,000 webcams to launch a distributed denial-of-service attack (DDoS) to bring down servers of large internet companies in 2016 [13]. In 2015, white hat hackers demonstrated the capability to assume full remote control of a Jeep to highlight the safety-critical nature of connected vehicle security risks [14].

Security of CPS is multilayered, as it involves the physical security of the system, security of the communication at each of the layers in the communication stack, storage security, and secure processing. Security solutions therefore need to holistically consider the system at all levels.

1.3.3 Privacy

The collection of physical data by CPS creates privacy concerns, at both the individual and organizational levels. For individuals, the high penetration of sensing devices, including smart phones and wearables, creates a digital biography of their lives and activities, which involves risks of the data being used by malicious entities or by commercial entities for monetary gain. Smartphone use, for instance, reveals a person's location with high spatial and temporal resolution, in addition to their activities that are captured by the phone's on-board sensors. Smart home sensors for security, smoke detection, and, more recently, voice activated interaction all record highly sensitive personal data that is vulnerable to exploitation. Smart meters readily reveal a person's activities as can be inferred and analyzed by the high-resolution energy usage data, which can lead to exploitation for break-ins into a person's home at time when the individual is not home.

At the organization level, sensing devices are being increasingly used for operational efficiency and traceability, which can lead to leaks in commercially sensitive information. In supply chains, for instance, sensors provide high-resolution information on the position and condition of traded goods. However, the supply chain participants may not want to reveal all their operational data to competitors within the system.

1.3.4 Heterogeneity in Device Resources

CPS encompasses a wide range of devices, from tiny passive radio frequency identification devices, active sensing devices, and edge networks to robotic and autonomous systems. The breadth of CPS devices leads to a high level of heterogeneity in device capabilities and computational resources, where cheaper and less capable devices can be deployed at scale for greater and denser spatial coverage, while higher-end platforms can include high-end processors, mobility, and autonomy for achieving more complex tasks. While the device heterogeneity in CPS drives the design of efficient and scalable systems to fit target applications, it poses challenges to the design of mechanisms that deliver security, privacy, trust, and auditability across the range of diverse devices. For instance, suitable mechanisms for an autonomous robot need to provide the highest security guarantees due to the safety-critical nature of operations, but these mechanisms may not suit lower-end CPS devices due to computational complexity and their inability to scale. The ability to design sufficiently versatile and configurable mechanisms to support the diverse landscape of CPS devices is therefore an open challenge.

1.3.5 Lack of Control or Auditability over Data

CPS data collection is typically automated so that sensing devices gather data from their physical environment on a periodic or event-triggered basis. As these technologies have become pervasive, their disappearance into the fabric of daily lives essentially strips users from the ability to control what data is collected and when it is collected. Automated data collection in CPS is also opaque where it is virtually impossible to audit the history of collected data, even by the data owners and authorized stakeholders. This limits the ability of stakeholder to extract value from the data, such as for process optimization or operational efficiency. Only limited approaches currently exist for providing control and auditability over CPS-generated data, which remains an open challenge.

1.3.6 Persistence/Sustainability

A CPS must operate for long durations without the need for human intervention in order to be viable. This requires high-quality software and hardware for CPS devices. It also requires the systems to be energy-efficient so that they do not deplete their energy budget. A CPS device's energy budget is subject to its battery capacity, and where applicable to the energy it can harvest from its environment. Without energy harvesting, CPS devices must optimize their operation and adapt to their environments in order to deliver the maximize utility. When energy harvesting is available, CPS devices have a dynamic energy budget that is subject to their physical context, namely, its current and prospective energy [2].

The persistence of a CPS is critical for serving its primary purpose. A key design challenge for CPS is therefore how to deliver maximum utility within its energy constraints. These constraints also need to be considered when designing solutions for other CPS challenges, such as privacy, security, or trust.

1.3.7 Trust

Trust is defined as the probability that a particular claim is true. In the context of CPS, trust refers to the likelihood that the data, processing, and decisions are genuine and uncompromised. Trusting CPS is particularly challenging as the data is typically an observation of the physical context, where the process of obtaining the physical data is vulnerable to attacks and malfunctions. Another trust challenge for CPS stems from decentralization, where multiple coordinating nodes must trust each other despite the possibility of malicious behavior by one or more of the participating nodes, due to being compromised by an attacker or due to selfish interests. A third trust challenge for CPS lies in verifying the identity of the participating nodes themselves. Recall that CPS nodes are embedded in their physical contexts, and any intended decoupling of a CPS node with its physical context may be undetected. This decoupling then leads to a mismatch between the node's digital identity and its represented physical identity. Consider for instance a sensing tag used for supply chain traceability. If the tag is moved from its original product to another product, it maintains it digital identity, yet its physical identity, namely, the product to which it is attached, has changed. This requires identity verification mechanisms that are the focus of researchers. In some instances, it is sufficient to verify the type of a CPS node without verifying its identity, to preserve its anonymity. For instance, a verified smart vehicle may contribute data to an intelligent transport network without revealing its identity. Anonymous identity trust is therefore a further open challenge.

1.4 Conclusion

This chapter has introduced CPS, which integrate sensing, networking, and computation capabilities where millions of networkable devices sense the environment to capture information. Using network infrastructure, mainly the internet, the information is transferred to cloud servers where it is processed to offer personalized services to the users. The data processing outcome is fed into decision-making algorithms that may trigger particular actions by the devices. CPS are applied in broad range of applications including but not limited to supply chain, smart grid, and smart city. CPS faces multiple challenges including centralization, security, privacy, heterogeneity of devices, trust, sustainability, and access control.

This book is structured into three parts: Part I includes Chapters 1 to 3, which discuss relevant background information. We study the existing solutions to the CPS challenges in Chapter 2 and introduce blockchain technology in Chapters 3. Part II contains Chapters 4 to 7 and elaborates on optimized blockchain frameworks for CPS applications. Part III includes Chapters 8 to 11 and outlines applications of blockchain in CPS. Finally, we conclude the book and outline future trends in Chapter 12.

References

[1] Rana, R., et al., "Feasibility Analysis of Using Humidex as an Indoor Thermal Comfort Predictor," *Energy and Buildings*, Vol. 64, September 2013, pp. 17–25.

[2] Purdon, S., et al., "Model-Free HVAC Control Using Occupant Feedback," *Proceedings of IEEE International Workshop on Global Trends in Smart Cities (GoSmart), co-located with IEEE International Conference on Local Computer Networks (LCN)*, Sydney, Australia, October 2013.

[3] Schwab, K., *The Fourth Industrial Revolution,* December 12, 2015, https://books.google.com.au/books?hl=en&lr=&id=ST_FDAAAQBAJ&oi=fnd&pg=PR7&dq=The+Fourth+Industrial+Revolution&ots=DTox6Tuw_K&sig=qv1wf0fuXGoYlXnx4twhf8J4HcE&redir_esc=y#v=onepage&q=The%20Fourth%20Industrial%20Revolution&f=false. Accessed January 15, 2019.

[4] Elijah, O., et al., "An Overview of Internet of Things (IoT) and Data Analytics in Agriculture: Benefits and Challenges," *IEEE Internet of Things Journal*, Vol. 5, No. 5, 2018, pp. 3758–3773.

[5] Suraki, M. Y., and M. Jahanshahi, "Internet of Things and Its Benefits to Improve Service Delivery in Public Health Approach," *2013 IEEE 7th International Conference on Application of Information and Communication Technologies*, October 2013, pp. 1–4.

[6] Lu, Y., and J. Cecil, "An Internet of Things (IoT)-Based Collaborative Framework for Advanced Manufacturing," *The International Journal of Advanced Manufacturing Technology*, Vol. 84, No. 5-8, 2016, pp. 1141–1152.

[7] Verdouw, C. N., et al., "Virtualization of Food Supply Chains with the Internet of Things," *Journal of Food Engineering*, Vol. 176, 2016, pp. 128–136.

[8] Shin, D. H., and Y. Jin Park, "Understanding the Internet of Things Ecosystem: Multi-Level Analysis of Users, Society, and Ecology," *Digital Policy, Regulation and Governance*, Vol. 19, No. 1, 2017, pp. 77–100.

[9] Reka, S. S., and T. Dragicevic, "Future Effectual Role of Energy Delivery: A Comprehensive Review of Internet of Things and Smart Grid," *Renewable and Sustainable Energy Reviews*, Vol. 91, 2018, pp. 90–108.

[10] Shi, J., et al., "A Survey of Cyber-Physical Systems," *2011 IEEE International Conference on Wireless Communications and Signal Processing (WCSP)*, November 2011, pp. 1–6.

[11] Hu, F., *Cyber-Physical Systems: Integrated Computing and Engineering Design*, Boca Raton, FL: CRC Press, 2013.

[12] Autonomous vehicle data generation, https://www.networkworld.com/article/3147892/one-autonomous-car-will-use-4000-gb-of-dataday.html.

[13] Mirai Botnet, https://www.csoonline.com/article/3258748/the-mirai-botnet-explained-how-teen-scammers-and-cctv-cameras-almost-brought-down-the-internet.

[14] Miller, C., and C. Valasek, "Remote Exploitation of an Unaltered Passenger Vehicle," *Black Hat USA*, Vol. 91, 2015.

[15] Geissdoerfer, K., et al., "Getting More Out of Solar-Harvesting Systems: Energy Management Under Time-Varying Utility with PREACT," *Proceedings of the International Conference on Information Processing in Sensor Networks (IPSN 2019)*, Montreal, Canada, April 2019.

2

Distributed Solutions for CPS

2.1 Introduction

CPS encompasses a broad range of heterogeneous devices that sense the environment and share information with service providers (SPs) to offer personalized automated services to the end-users, which enhances user satisfaction. Conventionally, CPS devices communicate through central controllers, resulting in a many-to-one network topology that suffers from the single-point-of-failure challenge. The number of devices connected in CPS and the range of services offered by SPs is continuously growing, which amplifies the associated centralization challenges and limits scalability. The SPs receive a large volume of personalized data about the users which enables them to build virtual profile of the users thus potentially compromising their privacy. To address these challenges, several approaches have been proposed to decentralize information flow in CPS. Decentralization requires solving new challenges, such as security, storage, and trust.

In this chapter, we provide a state-of-the-art review of the existing distributed frameworks for CPS. Figure 2.1 presents a taxonomy of the relevant CPS distributed solutions, in relation to addressing the key requirements of processing, communication, storage, and energy. Embedded processing and learning are aimed at achieving scalability and privacy by allocating processing tasks to various CPS nodes. Key considerations for communication in CPS include security, privacy, and trust. Several distributed mechanisms support CPS security, privacy, and trust, ranging from distributed authentication, distributed access control, and distributed key management, and trust and reputation

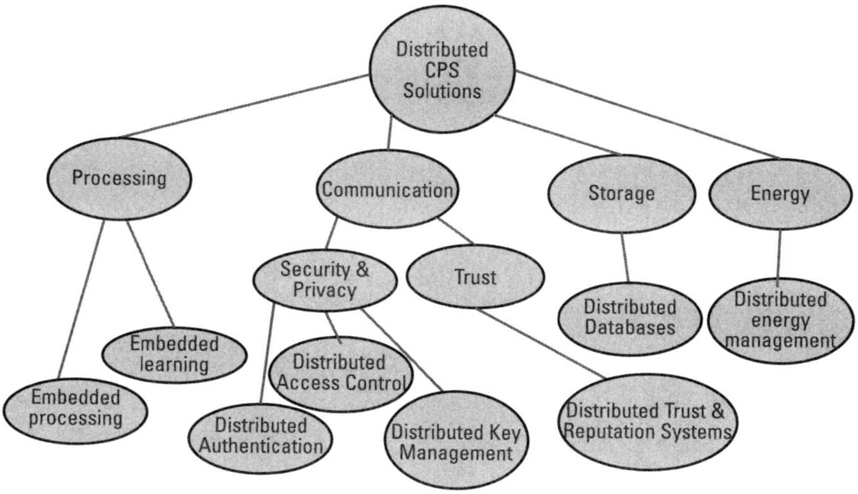

Figure 2.1 A taxonomy of distributed CPS solutions.

management systems, which we discuss in more detail later in this chapter. Storage is mainly supported by distributed databases, while distributed energy management algorithms for CPS/IoT are aimed at promoting system persistence.

In Chapter 1, we defined the seven challenges of CPS systems, namely, centralization, security, privacy, heterogeneity, auditability, persistence, and trust. Table 2.1 maps existing distributed CPS solutions to these challenges. We observe that while all distributed solutions aim to move away from centralization, some of them continue to rely on a trusted third party (TTP) for coordinating key CPS functions. While this supports trust in CPS interactions, it fails to achieve the ultimate goal of full decentralization. The only exception is trust and reputation systems, which explicitly represent and track trust in a distributed manner and therefore tackle both the trust and centralization challenges. There is also an interesting trade-off between privacy and trust, where we find that solutions that support trust often require explicit identification and information-sharing among participants, which may compromise their privacy. All but one, namely, distributed processing, address the heterogeneity and security requirements to some extent. Distributed energy management algorithms often address the requirement of persistence but typically do not deliver on other key requirements, such as security, privacy, trust, or auditability. In summary, we find that existing distributed CPS solutions are not able to concurrently address all the key CPS challenges.

In the following sections, we discuss each of the distributed CPS solution types in more detail. We first briefly cover distributed processing approaches for CPS, which range from embedded processing to lightweight machine learning approaches on devices or at the edge nodes, in Section 2.2. We then turn our

Table 2.1
Mapping CPS Challenges to Distributed Solutions

	Decentralization	Security	Privacy	Heterogeneity	Auditability	Persistence	Trust
Distributed Processing	√	—	√	—	—	—	—
Distributed Authentication	—	√	—	√	—	—	√
Distributed Access Control	√	√	√	√	—	—	
Distributed Key Management	—	—	—	√	—	—	√
Distributed Trust and Reputation Systems	√	√	—	√	—	—	√
Distributed Databases	√	√	—	√	√	—	—
Distributed Energy Management	√	—	—	√	—	√	—

attention to distributed communication mechanisms for CPS that include security, privacy, and trust in Section 2.3 and discuss these mechanisms at length as they motivate the core focus of this book in exploring blockchain for CPS. Next, we discuss distributed storage for CPS mainly in the form of distributed databases in Section 2.4, followed by an overview of distributed energy management in CPS in Section 2.5. Section 2.6 outlines the limitations of the state of the art, and Section 2.7 concludes the chapter by summarizing how the open challenges identified in this section motivate the consideration of blockchain for CPS.

2.2 Distributed Processing in CPS

CPS devices sense, think, and act in real time as they interact with their physical surroundings. While these devices are typically connected through wireless interfaces, they have limited communication bandwidth and energy capacity to transmit all their data to a central server for processing. These constraints have motivated several proposals into processing CPS data in situ at the embedded device level or at edge devices. More recently, approaches for running lightweight machine learning approaches on CPS devices have been proposed. We discuss embedded processing and embedded learning separately in more detail below.

2.2.1 Embedded Processing

Embedded processing involves with running computational tasks on a CPS device to process its acquired data. While embedded processing increases the processing overhead at the CPS device, it reduces the communication overhead of sending high-frequency raw data. Embedded processing also can implicitly contribute to privacy in sending processed rather than raw data, which results in information loss and contributes to partial data obfuscation.

Simple forms of embedded processing include data aggregation, where CPS devices may combine several sensor readings across space or time and report the aggregated result. More advanced forms of embedded processing include feature extraction, where CPS devices compute more diverse statistical features from the raw data and share these features with the edge devices and the cloud server. For instance, several works have conducted trajectory compression on high-frequency GPS traces obtained by a CPS node [1, 2], sending only trajectory segments rather than raw location data, in order to reduce the communication overhead while providing utility to the tracking application. CPS nodes can also use dead reckoning and estimate the expected information gain from a future sensor sample to guide GPS sampling decisions, as in [3].

More advanced embedded processing algorithms have relied on coordination among multiple CPS nodes to deliver on a common application goal while preserving their energy resources. For instance, Jurdak et al. [4] used a collaborative localization approach among mobile nodes to reduce GPS sampling frequency while maintaining a target location accuracy for a cattle-tracking application.

Embedded processing decisions can be rules-based or model-based. Rules-based approaches rely on a set of static preset rules that determine actions based on observed context. Model-based approaches use a model that attempts to capture the behavior of the physical world and guide CPS node actions. Models can be static, based on domain knowledge or previously available data describing the deployment context, or dynamic, in that they evolve over time according to changing context. This moves processing algorithms towards learning algorithms, which we discuss next.

2.2.2 Embedded Learning

Embedded learning deals with running machine learning algorithms on one or more CPS devices, potentially in coordination with edge devices. Machine learning typically involves training of models to infer patterns of the input/output data. When these models are trained prior to system deployment, this is referred to as offline learning. In contrast, online learning is when the models are continuously updated based on changing physical context. Offline learning is more computationally efficient for resource-constrained CPS devices, as they only need to run new input data through the pre-trained models in real time. However, offline learning is less versatile, as trained models are static and do not adapt to changing physical contexts and dynamics. Online learning is more computationally demanding for CPS devices, as the learning needs to occur during deployment. This provides online learning with higher versatility in adapting to changing physical contexts that may not be foreseeable at deployment time. The high versatility of online learning has made it attractive for CPS systems. For instance, Valencia et al. demonstrated online learning through an in situ genetic programming framework for IoT devices [5], and they followed up with a distributed version of online learning on embedded devices where the devices share their learned models periodically through the island model to increase their genetic diversity [6]. More recently, federated learning has been proposed by Google to distributed learning tasks across several distributed nodes, which contributes both to decentralization and to privacy [7].

Overall, distributed processing approaches contribute significantly to alleviate the challenges of centralization in typical stove-piped architectures. This decentralization typically comes at the cost of increased energy consumption for coordinating among the agents, which limits the persistence potential of such approaches. Distributed processing can contribute to privacy as it shares the outcomes of data processing rather than the data itself, which limits the amount of shared information. This privacy benefit is typically implicit and it is difficult to know what the privacy implications specific to the user of each algorithm. Most distributed processing approaches also do not incorporate security and trust mechanisms. To understand these aspects further, we focus next on the distributed communication approaches for CPS.

2.3 Distributed Communication in CPS

CPS agents communicate with each other to achieve the common system goals. While the communication among CPS agents shares many features with broader communication networks, it also involves specific challenges relating to the cyberphysical interface. For instance, the large scale of CPS networks and the resource constraints of their agents require highly decentralized and scalable mechanisms to provide security, privacy, and trust across these networks. We discuss these requirements further below.

2.3.1 Security and Privacy in CPS

In this section, we discuss the existing distributed solutions for CPS security and privacy that cover distributed authentication mechanisms (Section 2.3.1.1), distributed access control mechanisms (Section 2.3.1.2), and distributed key management mechanisms (Section 2.3.1.3).

2.3.1.1 Distributed Authentication Mechanisms

Authentication refers to proving the identity of a user or device to other parties within a network. Due to the ever-increasing degree of connectivity in CPS, it is critical that the nodes authenticate each other before communication. This ensures protection against malicious nodes that may impersonate another node. A distributed authentication framework should satisfy the following conditions [8]:

- The framework must not be controlled by a central authority.
- The protocol should incur small overhead on the participating nodes.
- The protocol must consider how to manage keys.
- The protocol must provide secure end-to-end communication and be able to detect attacks.

The authors in [9] proposed an anonymous authentication method that does not require the presence of trusted central authorities. The framework enables a group of participants to jointly generate a private key, where each participant maintains a share of the key but is not able to construct the comprehensive private key of the group. The group key can only be constructed when at least t participants (out of m, where $t < m$) share their part of the private key. Resource-consuming tasks are allocated to *super nodes,* which are typically resource-capable devices. The users are authenticated based on two attributes that are derived from a standard certificate issued by a trusted authority and are only known to the user. The attribute-based authentication enables the SPs to enforce attribute-based access control where only nodes with particular attributes can access resources. The user shares their attributes with a trusted central authority, who then issues an anonymous access credential (ACC). The ACC functions as a proof of knowledge proving that the ACC owner knows the attributes necessary to authenticate a user in the network.

In [10], the authors proposed a distributed authentication method that utilizes encryption algorithms to secure multiple steps in user authentication. The authentication happens in two phases. The first phase is the registration phase. In this phase, the requester sends a registration request to a trusted certificate authority that validates the identity of the requester. It is assumed that the certificate authority and the requester have a secure connection. If the requester is validated, a certificate is issued that is used by the requester in the next phase, the authentication phase. In this phase, the requester sends a request to access resources from the requestee. The requester first sends a hello message along with its certificate. If the certificate is valid, the requestee sends a hello message back to the requester. Next, the requester and the requestee collaboratively generate a session key used for their communications.

The authors in [8] proposed a distributed authentication framework for a smart grid environment. The proposed framework builds a secure layer on top of the communication layers to enhance the security of the communications. The proposed framework consists of three tiers: (1) central cloud servers, (2) distributed cloud servers, and (3) smart meter (SM), energy providers (EP), and gateways (GW).

The Trusted Authority (TA) in each layer generates a public and private key pair and announces its PK to the nodes connecting to it in the lower layer. Each TA runs a key generation center (KGC) that generates partial private key (PPP) and partial public key (PPU) and distributes to the nodes connected to the TA. The nodes utilize the received PPP and PPU to build their own key pairs, which hides the PK of each user from the corresponding TA and thus enhances the privacy of the users. Assume that a SM and a GW wish to establish a connection and thus need to authenticate each other. First, the SM retrieves the PK of the GW from its corresponding TA. Then the SM generates a message containing its own PK with the PK of the GW. By receiving the message, the GW can decrypt the message using its private key and access the PK of the SM. The GW verifies the received PK by sending a request to its corresponding TA. If the PK is verified, the GW can trust that a valid device that is registered with a TA is generating the message. The GW then sends a message to the SM indicating that it accepts the connection. This enables the GW and SM to authenticate each other while protecting their privacy.

It is worth noting that while these approaches achieve decentralized authentication, their reliance on a trusted third party for issuing ACC means that they are still partially centralized. Once nodes are authenticated, other security services can be decentralized, such as access control, which we discuss next.

2.3.1.2 Distributed Access Control Mechanisms

Access control refers to monitoring and controlling the access to resources in CPS and is a way to enforce security and privacy. Distributed access control mechanisms vary in terms of the topology of access control from hierarchical, where designated nodes manage access, to individual, where nodes independently manage access, and peer-to-peer, where nodes jointly manage access.

The authors in [11] proposed a decentralized hierarchical access control framework. Each hierarchy is a set of locations where security policies are applied. In each hierarchy, a reference monitor node functions as the leader of the hierarchy that connects the members of the hierarchy to other hierarchies. The hierarchy members send their packets to the reference monitor to be sent to the destination (which can potentially be in another hierarchy or in the same hierarchy). The reference monitor is selected in a distributed manner and is responsible to enforce the policies defined at the hierarchy by monitoring the incoming and outgoing packets.

The authors in [12] proposed a distributed access control framework where each device individually controls access requests to its data. The owner of a device initially generates a token capability that authorizes the device (say, D_r) to access specific devices or data. This is considered the first access control layer where even the owner of the requester device must authorize the device for such a request, which enhances the security of the framework. D_r then generates an access request packet that contains the token issued by the owner and the signature of the device. The request is sent to the destination device (say, D_t), that is, the device that D_r is going to access. On receipt of the packet, D_t verifies the signature of D_r. If valid, D_t authorizes the access based on its own token capabilities. D_t may also seek permission from the owner. The result is then communicated back to the D_r.

The authors in [13] proposed a distributed privacy-preserving access control framework for wireless sensor networks (WSN) where nodes jointly manage access through a ring signature. The participants in the network can be categorized into three main groups: sensors, users, and the network administration. The network administration is a central trusted authority that is responsible for grouping the network in different clusters during network bootstrapping. The users in each group have the same access level to the sensors. To protect the privacy of the users, a single ring signature is used in each cluster. The ring signature enables a group of nodes to sign a message while the message verifier can verify that the message is signed by a genuine node but cannot link it to a single user in the group, which increases user privacy. To request data from a particular sensor, the user signs the request with the private key associated with the group key. The sensor can authorize the user by verifying the ring signature of the received packet.

The authors in [14] proposed a distributed access control framework for cloud storages that uses a peer-to-peer network to distribute the access keys. Key distribution centers (KDC) join the peer-to-peer network and distribute keys between the users. Each user is associated with a set of permissions (i.e., attributes) that collectively form a permission tree for the user. In the latter, the leaves are the attributes and the parents are Booleans of AND/OR. The KDCs distribute the keys based on the user permissions and their corresponding attribute trees. The users may request to gain particular attributes from the KDCs, which are verified and assigned by the authorities. Using the received key from KDC, the users can encrypt data in the cloud storage. The attribute-based encryption enables users to share data with a group of nodes using a single key.

By employing access control mechanisms, the users can authorize other participants to access their data. To enhance the security of the communications and the associated data, these are encrypted, which raises key management challenge, which is studied in the next section.

2.3.1.3 Distributed Key Management

Asymmetric encryption is used to secure communications between CPS participants that employs a pair of public and private keys to encrypt the content and thus protect against malicious nodes that may attempt to read a message, and sign the hash of the message to protect the integrity of messages and protect against non-repudiation attacks where a node may deny participating in a communication. Utilizing asymmetric encryption, the nodes must securely store the private key which complicates key management in CPS. Additionally, the CPS participants may employ multiple encryption keys to communicate with other nodes to enhance their privacy [15], which creates a challenge for keeping track of the keys used to communicate with each node. The outlined limitations make key management challenging. Conventional methods in the literature rely on centralized nodes to manage the keys and distribute keys among the CPS nodes; however, such methods suffer lack of security, privacy, and scalability, which has driven research into decentralized key management.

In [15], Hao et al. proposed a distributed key management framework for Vehicular Ad hoc Networks (VANET) where the vehicles communicate to exchange traffic data. In the proposed framework the keys are distributed by the Road Side Infrastructure (RSI). Each RSI distributes keys to the vehicles that are connected to that RSI and thus are in close vicinity. To reduce the overheads in managing the keys, each RSI groups its participants and distributes a group key between them. The group keys are generated by authorities (e.g., smart city managers) and are securely transferred to the RSIs. When an RSI receives a join request from a vehicle that wants to communicate with other vehicles joining the same RSI, it distributes a group private key to the vehicle, which can be used to sign the exchanged messages. Using the signature, the vehicles can authenticate each other and confirm if a vehicle belongs to the same RSI.

In [16], Lu et al. proposed a distributed key management framework for WSN considering the heterogeneity of the sensors and in particular different wireless range of the sensors. Initially, the sensors are grouped in I different groups based on the communication range, battery level, and available computational resources of the sensors. During the network bootstrapping phase, a central server distributes polynomial shares to the sensors. Later, if two sensors need to establish keys, they directly communicate and exchange information to share a key pair. To establish keys, the framework utilizes a randomly generated pool of bivariate polynomials.

In [17], Adjih et al. proposed a distributed key management framework to enhance the security of the routing algorithms in Mobile Ad hoc Networks (MANET). A central trusted PK infrastructure (PKI) assigns or certifies the keys generated by the participating nodes. All participants know the PK of the PKI and thus can verify its signature. The nodes utilize a certificate issued by the PKI to communicate with other nodes in the network.

In summary, distributed key management is a fundamental challenge in securing CPS. Most of the existing distributed solutions rely on trusted authorities, which may lead to centralization and highlights the demand for purely distributed key management frameworks. While the distributed security and privacy approaches are effective, they do not guarantee the reliability or trustworthiness of the distributed participants. This has fueled research into distributed trust and reputation systems for CPS, which we discuss next.

2.3.2 Trust and Reputation Systems for CPS

The lack of a central controller in distributed systems raises the challenge of lack of trust between the participants where the participating nodes may not potentially trust the generator of a message or trust that the other side of the communication will commit to their obligations. In a centralized system, the central controller validates all messages and thus functions as the trusted entity, which also establishes trust between the participants. In this section, we study the literature on distributed trust methods.

In [18], the authors proposed a distributed trust model to establish trust between the participants in peer-to-peer networks. The proposed trust model relies on direct and recommended trust, defined as follows:

- *Direct trust:* Node A has direct trust about node B if it has previously received and validated a message from node B.
- *Recommended trust:* Node A has recommended trust about node C if it receives a recommendation from node B about the trustworthiness of node C. It is assumed that node A has direct trust about node B and accepts recommendations from node B.

Each node in the network maintains a trust table to store the trust value for other nodes in the network. Direct and recommended trust values for each node are chosen from the range of [1, 4] where −1 is completely untrusted and 4 represents a fully trusted node. The participating nodes change the trust rating of a node based on the behavior of the nodes and the packets generated by them.

Similarly, the authors in [19] proposed a trust framework that utilizes the information gathered by the node and the recommendations received from other nodes to measure trust of the nodes over a distributed file-sharing platform. Initially, all nodes are ranked as not trusted and each node only maintains the trust rating for its one-hop neighbors. When node A receives data from node B, it verifies the data by matching the data with the data received from other peers in the vicinity. If the data is verified, the trust rating for node B is increased and node A attaches a recommendation to the packet and sends to its neighbors. The nodes that receive the packet with the recommendation must first attempt to verify the data without relying on the recommendation. If the node is unable to verify the data, it evaluates the recommendation. The weight of a recommendation depends on the trustworthiness and the level of confidence of the recommender.

In [20], the authors proposed a distributed trust and reputation framework for MANET. Each node in the network listens to the communications of all its neighbors to detect misbehavior activities. Each node stores a local copy of the packets sent to its neighbors, eavesdrops the packets sent by the neighbors, and compares them with the local copy. If any change is detected, the neighbor node is marked as malicious. In each node, a trust manager entity calculates the trust rate and observes the behavior of the neighbor nodes. Once the misbehavior of a node is detected, the trust manager generates an *alarm* packet and informs other nodes in the network of the identity of the malicious node. The proposed framework employs trust to choose the best path toward a destination that is managed by a path manager in each node. The routing protocol detects multiple paths toward a particular destination and sends those to the path manager. The latter then evaluates the trust rate of all participating nodes in each path and calculates a trust rate accordingly. The path with the highest trust rate is selected as the best path, which increases the security of the framework against malicious activities.

Overall, distributed communication approaches may be able to achieve security and privacy in CPS and establish trust and reputation among the participants. However, most of the existing approaches offer limited services (e.g., only offer access control) and increase computational, memory, and packet overheads. CPS devices exchange large volume of data stored in a database accessible by an SP to offer personalized services to the users. Next, we discuss distributed storage solutions for CPS.

2.4 Distributed Storage in CPS

The CPS devices generate a large volume of data, which is typically stored in a database to be shared with other participants or service providers. However, storing such large volume of data in a central cloud storage does not scale and creates a single point of failure. Also, the central database remains an attractive target for malicious entities to gather information of the devices and thus compromise the security and privacy of the CPS users. As the volume of data stored in a database increases, the processing overhead and the associated delay in querying content increases, which impacts the cost in managing the database. Distributed databases are proposed in the literature to address these limitations.

In this section, we review some of the existing solutions. The design of distributed databases involves the following challenges [21]:

- *Database distribution:* This refers to how to distribute the database across different sites. There are two main methods to place the database: (1) partitioning where data is partitioned and shared across multiple nodes, and (2) replication where the same version of the data is shared between multiple nodes.
- *Database directory:* A directory identifies the location where the files are stored in the distributed database. Directories can be global to include all different partitions in the database or local for each part of the database.
- *Database querying:* This refers to how to query and retrieve particular data in the database.
- *Database concurrency control:* This refers to synchronization of data among multiple sites.
- *Database reliability and recovery:* This refers to ensuring that data is always available to the genuine nodes that require access. In case of a fault in the network, the framework should be able to recover the data and ensure availability.

The Inter-Planetary File System (IPFS) [22] is a distributed database that allows the users to store and revoke data. Each user can be a client and a server at the same time. In the client mode, the user downloads data from other users in the network, while in the server mode, other users can download data from the user, which enables the distributed storage of data. IPFS includes three core components:

1. *Content-based addressing:* Conventional file addressing methods use the file name and the address of the storage where the data is stored to identify the data, which potentially leads to duplicate data. It also makes it difficult to locate files, particularly in distributed file-sharing platforms where data is shared across multiple nodes. To reduce the associated overheads, IPFS uses the hash of the data as the address of that particular data, which introduces content-based addressing.
2. *Directed Acyclic Graph (DAG):* DAG is a type of graph without any loops as shown in Figure 2.2. When a user decides to store a file in IPFS, the file is divided into multiple parts. This potentially reduces the overhead in updating the file content. Assume that a user em-

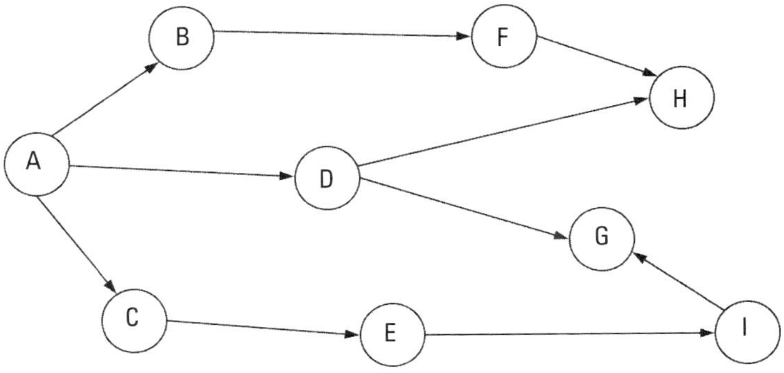

Figure 2.2 An example of a DAG.

ploys IPFS to store his or her website data. The user then decides to change the photo in the main website page. By storing data in different parts, the user only needs to change the photo without the need to update the content of the entire website. Additionally, the data might be stored in a file and that file, along with multiple other files, might be placed in a folder, and the folder might be in another folder and so on. IPFS uses a Merkle-DAG where the parts of data are considered as leaves of a binary tree and form a tree. The tree is employed to search for a particular chunk of data.

3. *Distributed Hash Table (DHT):* DHT enables the users to locate a particular chunk of data. In a DHT, an example of which is shown in Figure 2.3, the data is stored in a key-value pair where each peer in the DHT stores data mapped to that peer. DHT enables peers to leave or new peers to join with a small overhead. To retrieve the data, a node sends its request to the closest peer in the DHT. The request is then forwarded to reach the corresponding peer. As an example, consider the DHT shown in Figure 2.3. Assume that node A is looking for content address "aclkajiero129ahf2" and only knows about node 4. Node A sends a request to node 4. Based on the DHT, node 4 is only responsible for data for which the address starts with [A–K]. Thus, node 4 forwards the request to node 5. The steps are repeated until the request reaches node 2 that stores the relevant content. Node 2 sends a response back to node A and thus node A can start to download the content. IPFS employs DHT to look up the peer corresponding to a particular piece of data.

In IPFS, the data of a user is stored in a number of other nodes; however, there is no guarantee that the data is always available to the user as the nodes

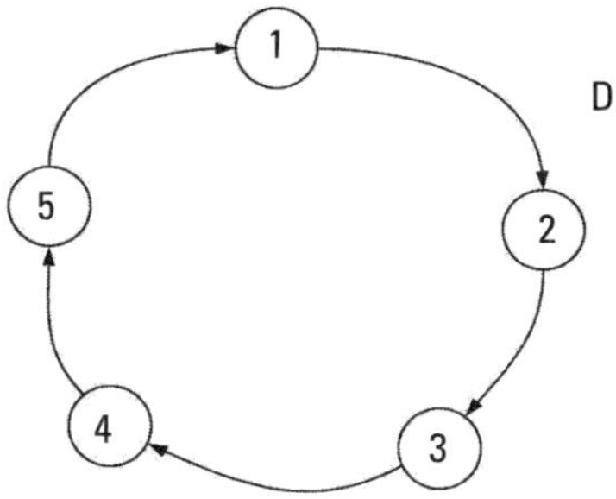

Distributed Hash Table (DHT)

Node	Key
1	0-9
2	a-k
3	l-z
4	A-K
5	L-Z

Figure 2.3 An example of a DHT.

that have stored the data may leave the network. A user may pin particular data by storing data permanently in their local machine; however, even if data is pinned by a user, when the user leaves the network, the data is lost.

BitTorrent [23] is another example of a distributed file-sharing method. To request a file, the user (say, peer A) first broadcasts a packet that contains the type of the data that it requires. The peer that has stored the requested data (say, peer B) sends a response back to peer A. Peer A then starts downloading the file from peer B. Broadcasting a packet to locate the node that has stored a particular data incurs significant packet overhead and limits scalability. To address this challenge, BitTorrent employs a tracker server that is responsible for finding the nodes with a particular data and returning a swarm of nodes to the user. The swarm is essentially the group of nodes that have full or part of the requested data. While peer A is downloading a file, other users start downloading from it, which enables distributed sharing of files and thus eliminates the need for central databases. This also protects against the nodes that only join the network to download a file and leave the network after downloading is concluded. To increase the speed of downloading files, BitTorrent enables parallel file downloading where a user can download from multiple users simultaneously. BitTorrent employs tit-for-tat protocol where a peer sends data with high speed to *k* top nodes that are sending data with high speed to it. Additionally, the node selects a random node in the network to send data with high speed that enables nodes that are not in the top 4 list to receive data with high speed that is known as opportunistic unchoking.

In summary, the distributed database solutions enable users to distribute their data among multiple parties. Data availability, data security, and privacy of the users remain open research challenges. Additionally, CPS devices have limited energy sources and are required to remain functional for a long period of time. This makes energy management challenging and this is discussed next.

2.5 Distributed Energy Management in CPS

CPS devices are embedded in their physical deployment environments, which require long-term unattended operation (i.e., persistence). CPS devices typically rely on energy storage through batteries or capacitors to cover their energy needs and, in some cases, on energy harvesting from the environment to replenish their energy [24]. There have been several works on minimizing the energy consumption [25] and managing the precious harvested energy [26] on the miniaturized sensor node.

Distributed energy algorithms are aimed at maximizing the operational lifetime of CPS node. They include transmission power control [27–29], scheduling at the medium access control (MAC) layer scheduling [30–32] and task scheduling [33–36]. Transmission power control enables CPS nodes to vary their transceiver transmission power, individually or in collaboration, in order to optimize multiple objectives, including communication utility and energy consumption. MAC layer scheduling, which can be time-based or adaptive to context, varies the access schedules of CPS nodes to the wireless medium according the operational constraints.

Task-scheduling algorithms take a broader view, with the ability to schedule tasks at different layers of the stack in CPS nodes, such as sensor sampling, processing tasks, and transmission. They are thus effective at minimizing the energy consumption on the CPS node.

Task-scheduling algorithms are more effective in reducing the energy consumption, than other communication-focused energy management schemes (i.e., transmission power control) due to their interaction with a broader task set (such as sampling, processing, and communication). A separate body of work involves adaptive sampling of sensors in CPS devices to manage energy, based on changing contexts and on dynamic energy availability [37].

Energy management algorithms in CPS prioritize the energy performance, typically maximizing utility subject to an energy constraint or minimizing energy consumption subject to a utility constraint. They contribute to the CPS goals of persistence (by design), decentralization (as they allow nodes to act independently or in groups), and heterogeneity (as algorithms can be easily tailored to the constraints of a single CPS node). However, their design priority of persistence typically sacrifices security, privacy, and trust goals. Having discussed the different distributed solution types for CPS, we focus on the open issues next.

2.6 Limitations and Open Questions

In this section, we outline the key limitations of the existing distributed solutions in CPS. As we saw in Table 2.1, existing distributed algorithms for CPS tackle some but not all of the CPS requirements. Let's revisit the key limitations here:

- *Reliance on TTP:* Most of the existing works in the literature rely on a TTP to conduct part or all the fundamental tasks in the network (e.g., a trusted CA to verify or distribute the PKs). While this has the benefit of providing trusted interaction among participants, it fails to satisfy the goal of decentralization, as the TTP effectively maintains central control. What is needed is an architecture that supports both decentralization and trust, or at least consensus, among potentially untrusted participants.
- *Lack of privacy:* Most of the existing works do not explicitly consider the privacy of the users or provide limited privacy. Studies have shown that the malicious nodes may compromise the privacy of the users by studying their packets even if the associated data is encrypted [38]. There is a fundamental trade-off between privacy and trust, where participants that reveal more information can be more trusted yet give away more of their privacy. There is therefore a need for approaches that can support trust and privacy, such as verification of assertions by other participants without forcing these participants to reveal their sensitive data.
- *Scalability and overheads:* Running the existing algorithms consumes significant computational, storage, bandwidth, and energy of the participating nodes, which is beyond the capabilities of most CPS devices. While energy management algorithms can solve these problems, they typically do not provide any security, privacy, or trust, which creates significant vulnerabilities for current persistent CPS networks. There is a need for lightweight approaches that consider the large scale and resource constraints of CPS networks, while ensuring security, privacy, and trust.
- *Limited services:* Most of the existing frameworks only offer a single or limited number of services to the users; for example, a distributed trust framework only enables the participants to establish trust. CPS demands a framework that can offer a broad range of services with limited overhead. We have seen in Table 2.1 that none of the existing distributed CPS solutions can deliver on all of the CPS requirements identified in Chapter 1. There is therefore a need for a new architecture that can better support these requirements.
- *Lack of transparency and auditability:* The auditability of the history of communications in the CPS enhances user privacy and system security as the users can monitor the interactions of their devices. In the existing

frameworks, the participating nodes have no or limited ability in auditing the history of interactions in the network, which limits the transparency of the systems. This raises the need for a distributed database that guarantees that the stored information does not change over time (i.e., is immutable) or at least that any malicious changes are easily detected.

The outlined challenges in the existing distributed solutions for CPS highlight the need for a new framework that offers a broad range of services while considering particular requirements of the IoT devices including resource restriction and heterogeneity of the devices. In the next chapter, we discuss blockchain as a novel solution that has the potential to address the outlined challenges.

2.7 Conclusions

CPS encompasses a broad range of resource-restricted heterogeneous devices that communicate and share data to offer personalized and automated services to the users. Conventional CPS frameworks rely on centralized brokered communications where all devices are connected, authenticated, and authorized through a central controller. Thus, even if two devices are sitting next to each other, they may have to communicate through the cloud server of the service providers. This potentially raises the security and privacy concerns and introduces single point of failure and many-to-one traffic, which leads to limited scalability. In recent years, distributed solutions have been proposed to address the outlined challenges. In this chapter, we studied the existing distributed security frameworks, databases, and trust and reputation frameworks. Although the existing solutions reach promising benefits, they suffer from challenges including reliance on TTPs, high overheads, lack of privacy, limited services, and lack of transparency.

In the next chapter, we will study the applicability of blockchain in CPS to address the outlined limitations. We will see how blockchain, as a distributed ledger technology, can address some of the key CPS requirements, such as decentralization, immutability, and security, by having multiple network nodes maintain a copy of the ledger. The ledger is structured into linked block that are linked through cryptographic hashes, such that any changes to previously stored information is easily detected by participating nodes. We will also see that while blockchain offers significant opportunities for CPS, its successful adoption for CPS requires consideration of the CPS challenges of scalability, storage overhead and compliance with the right to be forgotten regulation, privacy, and trust, which we discuss in Part II of the book.

References

[1] Murtaza, G., et al., "Trajectory Approximation for Resource Constrained Mobile Sensor Networks," *Proceedings of the 9th IEEE Conference on Distributed Computing in Sensor Systems (DCOSS)*, Marina Del Rey, CA, May 2014.

[2] Liu, J., et al., "Bounded Quadrant System: Error-Bounded Trajectory Compression on the Go," *Proceedings of the 31st IEEE International Conference on Data Engineering (ICDE)*, Seoul, South Korea, April 2015.

[3] Sommer, P., et al., "Information Bang for the Energy Buck: Energy- and Mobility-Aware Tracking," *Proceedings of the International Conference on Embedded Wireless Systems and Networks (EWSN)*, Graz, Austria, February 2016.

[4] Jurdak, R., et al., "Adaptive GPS Duty Cycling and Radio Ranging for Energy-Efficient Localization," *Proceedings of the 8th ACM Conference on Embedded Networked Sensor Systems (Sensys)*, Zurich, Switzerland, November 2010, pp. 57–70.

[5] Valencia, P., P. Lindsay, and R. Jurdak, "Distributed Genetic Evolution in WSN," *Proceedings of the 9th International Conference on Information Processing in Sensor Networks (IPSN)*, Stockholm, Sweden, April 2010.

[6] Valencia, P., et al., "Genetic Programming for Smart Phone Personalisation," *Applied Soft Computing*, Vol. 25, September 2014, pp. 86–96.

[7] Konecný, J., et al., "Federated Learning: Strategies for Improving Communication Efficiency," arXiv preprint arXiv:1610.05492, 2016.

[8] Saxena, N., and B. J. Choi, "Integrated Distributed Authentication Protocol for Smart Grid Communications," *IEEE Systems Journal*, Vol. 12, No. 3, 2016, pp. 2545–2556.

[9] Alcaide, A., et al., "Anonymous Authentication for Privacy-Preserving IoT Target-Driven Applications," *Computers & Security*, Vol. 37, 2013, pp. 111–123.

[10] Porambage, P., et al., "Two-Phase Authentication Protocol for Wireless Sensor Networks in Distributed IoT Applications," *2014 IEEE Wireless Communications and Networking Conference (WCNC)*, April 2014, pp. 2728–2733.

[11] Dell'Amico, M., et al., "Hipolds: A Hierarchical Security Policy Language for Distributed Systems," *Information Security Technical Report*, Vol. 17, No. 3, 2013, pp. 81–92.

[12] Skarmeta, A. F., et al., "A Decentralized Approach for Security and Privacy Challenges in the Internet of Things," *2014 IEEE World Forum on Internet of Things (WF-IoT)*, March 2014, pp. 67–72.

[13] He, D., et al., "Distributed Access Control with Privacy Support in Wireless Sensor Networks," *IEEE Transactions on Wireless Communications*, Vol. 10, No. 10, 2011, pp. 3472–3481.

[14] Ruj, S., A. Nayak, and I. Stojmenovic, "DACC: Distributed Access Control in Clouds," *2011 IEEE 10th International Conference on Trust, Security and Privacy in Computing and Communications*, November 2011, pp. 91–98.

[15] Hao, Y., et al., "A Distributed Key Management Framework with Cooperative Message Authentication in VANETs," *IEEE Journal on Selected Areas in Communications*, Vol. 29, No. 3, 2011, pp. 616–629.

[16] Lu, K., et al., "A Framework for a Distributed Key Management Scheme in Heterogeneous Wireless Sensor Networks," *IEEE Transactions on Wireless Communications*, Vol. 7, No. 2, 2008, pp. 639–647.

[17] Adjih, C., D. Raffo, and P. Muhlethaler, "Attacks Against OLSR: Distributed Key Management for Security," *2nd OLSR Interop/Workshop*, Palaiseau, France, Vol. 14, July 2005, pp. 1–5.

[18] Abdul-Rahman, A., and S. Hailes, "A Distributed Trust Model," *Proceedings of the 1997 Workshop on New Security Paradigms*, January 1998, pp. 48–60.

[19] Can, A. B., and B. Bhargava, "Sort: A Self-Organizing Trust Model for Peer-to-Peer Systems," *IEEE Transactions on Dependable and Secure Computing*, Vol. 10, No. 1, 2012, pp. 14–27.

[20] Buchegger, S., and J. Y. Le Boudec, "Performance Analysis of the CONFIDANT Protocol, *Proceedings of the 3rd ACM International Symposium on Mobile Ad Hoc Networking & Computing*, June 2002, pp. 226–236.

[21] Özsu, M. T., and P. Valduriez, *Principles of Distributed Database Systems, Vol. 2*, Englewood Cliffs, NJ: Prentice Hall, 1999.

[22] "IPFS File Distribution," https://ipfs.io/.

[23] BitTorrent, https://www.bittorrent.com.

[24] Sandhu, M., et al., "Towards Energy Positive Sensing Using Kinetic Energy Harvesters," *Proceedings of the 18th Annual IEEE International Conference on Pervasive Computing and Communications (PerCom)*, Austin, TX, March 2020.

[25] Jang, T., et al., "Circuit and System Designs of Ultra-Low Power Sensor Nodes with Illustration in a Miniaturized GNSS Logger for Position Tracking: Part I—Analog Circuit Techniques," *IEEE Transactions on Circuits and Systems I: Regular Papers*, Vol. 64, No. 9, 2017, pp. 2237–2249.

[26] Ejaz, W., et al., "Efficient Energy Management for the Internet of Things in Smart Cities," *IEEE Communications Magazine*, Vol. 55, No. 1, 2017, pp. 84–91.

[27] Wu, Y., et al., "Optimal Relay Selection and Power Control for Energy-Harvesting Wireless Relay Networks," *IEEE Transactions on Green Communications and Networking*, Vol. 2, No. 2, 2017, pp. 471–481.

[28] Lin, S., et al., "ATPC: Adaptive Transmission Power Control for Wireless Sensor Networks," *ACM Transactions on Sensor Networks (TOSN)*, Vol. 12, No. 1, 2016, pp. 1–31.

[29] Xiao, S., et al., "Transmission Power Control in Body Area Sensor Networks for Healthcare Monitoring," *IEEE Journal on Selected Areas in Communications*, Vol. 27, No. 1, 2009, pp. 37–48.

[30] Sherazi, H. H. R., L. A. Grieco, and G. Boggia, "A Comprehensive Review on Energy Harvesting MAC Protocols in WSNS: Challenges and Tradeoffs," *Ad Hoc Networks*, Vol. 71, 2018, pp. 117–134.

[31] Chang, C. -Y., and H. -R. Chang, "Energy-Aware Node Placement, Topology Control and MAC Scheduling for Wireless Sensor Networks," *Computer Networks*, Vol. 52, No. 11, 2008, pp. 2189–2204.

[32] Jang, B., J. B. Lim, and M. L. Sichitiu, "An Asynchronous Scheduled MAC Protocol for Wireless Sensor Networks," *Computer Networks*, Vol. 57, No. 1, 2013, pp. 85–98.

[33] Caruso, A., et al., "A Dynamic Programming Algorithm for High-Level Task Scheduling in Energy Harvesting IoT," *IEEE Internet of Things Journal*, Vol. 5, No. 3, 2018, pp. 2234–2248.

[34] Baghaee, S., et al., "Demonstration of Energy-Neutral Operation on a WSN," *European Wireless 2014,* 20th European Wireless Conference, pp. 1-6.

[35] Rault, T., A. Bouabdallah, and Y. Challal, "Energy Efficiency in Wireless Sensor Networks: A Top-Down Survey," *Computer Networks*, Vol. 67, 2014, pp. 104–122.

[36] Xiong, S., et al., "Multiple Task Scheduling for Low-Duty-Cycled Wireless Sensor Networks," *2011 Proceedings IEEE INFOCOM*, 2011, pp. 1323–1331.

[37] Alippi, C., et al., "An Adaptive Sampling Algorithm for Effective Energy Management in Wireless Sensor Networks with Energy-Hungry Sensors," *IEEE Transactions on Instrumentation and Measurement*, Vol. 59, No. 2, 2009, pp. 335–344.

[38] Apthorpe, N., D. Reisman, and N. Feamster, "A Smart Home Is No Castle: Privacy Vulnerabilities of Encrypted IoT Traffic," *arXiv preprint arXiv:1705.06805*, 2017.

3

Blockchain for CPS

3.1 Introduction

In recent years, blockchain, a distributed ledger of chained blocks, has attracted significant attention as a means to address the challenges in CPS including security, anonymity, centralization, trust, and auditability (see Chapter 1 for further detailed CPS challenges). Blockchain was introduced in Bitcoin [1], the first cryptocurrency, in 2008 as the underlying technology which enables participating nodes to exchange coins without relying on a central authority and thus builds a trusted network over untrusted participants. Since the introduction of Bitcoin, multiple other cryptocurrencies have been created with different purposes that are known as Altcoins [2].

Although blockchain was first applied to monetary applications, it has received tremendous attention from both academics and practitioners with a view towards using it in nonmonetary applications thanks to its salient features including security, anonymity, decentralization, trust, and immutability. The basic communication primitive between the blockchain participants is known as a transaction. The hash of the transaction content is signed by the transaction generator, which enhances the security of blockchain communications. The participating nodes employ a changeable public key (PK) as their identity, which introduces a high degree of anonymity by concealing the real-world identity of the users. Blockchains are managed in a distributed manner, where all transactions and blocks are verified by the participating nodes. Particular nodes in the network, known as miners or validators (for consistency, we refer to these nodes as validators in the rest of this book), may choose to append new blocks to the blockchain in return for some incentives, normally in the form of tokens or coins. Appending a new block involves following a consensus algorithm, whereby all validators must agree on the addition of the new block, which introduces high security and trust. Each block maintains a pointer to the previous block that is the hash of the previous block, which makes block modification impossible, rendering blockchains highly immutable.

The aforementioned features of blockchain make it a potential candidate to address security, trust, centralization, and auditability challenges in a broad range of applications in multiple disciplines including law, finance, and computer science. Blockchain can thus potentially address the challenges in CPS outlined in Chapter 1 or mitigate their impact. However, applying blockchain in CPS is not straightforward due to the specific features of CPS including the heterogeneity and limited resources of participating nodes, the large scale of CPS systems, and the real-time nature of these applications. In the rest of this chapter, we first discuss the basic blockchain concepts in Section 3.2. We then outline the challenges involved in adopting blockchain for CPS in Section 3.3. Section 3.4 concludes this chapter and summarizes key observations.

3.2 Blockchain

In this section, we provide a background on blockchain. Section 3.2.1 introduces the structure of blockchain and outlines its core concepts. Section 3.2.2 outlines the process of storing new blocks in the blockchain and Section 3.2.3 outlines widely used consensus algorithms employed in blockchains.

3.2.1 Blockchain Structure

Blockchain is a database that is shared across all participating nodes in a distributed network. The type of the stored data in the blockchain depends on the application. For example, in cryptocurrencies like Bitcoin the data is the history of the coin exchanges, while in a smart grid setting the data can be the energy usage, demand, or price.

Blockchain can be perceived as the combination of four known concepts in computer science: (1) linked lists, (2) digital signature, (3) hash functions, and (4) peer-to-peer networks. A linked list is a well-known data structure that chains data elements into a linear list, where new data is linked to the last added data in the list as shown in Figure 3.1 [3].

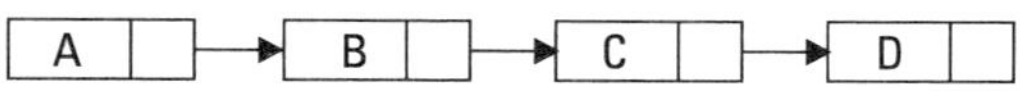

Figure 3.1 The structure of a linked list.

The second key element of blockchain is the digital signature, which is part of encryption algorithms. One of the main aims of encryption algorithms is to ensure that only authorized persons can read messages. Encryption methods are categorized as symmetric encryption, where the message can be encrypted and decrypted using the same key, also known as shared key, and asymmetric encryption, where separate keys are employed to encrypt and decrypt messages. Each node (say, Alice) maintains a pair of keys that constitutes of a public key (PK^+) and a private key (PK^-). A message encrypted with one key can only be decrypted with the other key, that is, if a message is encrypted by PK^+, it can only be decrypted using the corresponding PK^-. A message encrypted by the PK^- of Alice is known as its signature given that only Alice knows the PK^-. Other nodes can verify the signature, that is, decrypt the message, using the corresponding PK^+ of Alice.

The third element of blockchain is cryptographically secured hash function, which is a function that converts an input data of any size to a randomly generated output string with a fixed size. The fundamental features of a cryptographically secure hash function are:

- A small change in the input significantly changes the output.
- The size of the output is independent of the size of the input.
- It is not probabilistically feasible to find two messages with the same hash.
- It is not probabilistically feasible to find the input message just by knowing the hash function output.

In blockchains, the Secure Hash Algorithm (SHA) [4] family is employed, in particular, SHA-256, as the main hashing function.

The final element of blockchain is peer-to-peer networks where the participants broadcast packets among each other and there is not a central controller. The peer-to-peer network does not enforce any particular communication structure for the nodes.

The four elements outlined above are the core enablers of blockchain technology. Similar to linked lists, blocks of data are linked in blockchains where each block contains a number of transactions; however, blockchain uses a hash function output of one block to link it to the next block, effectively creating a chain instead of normal pointers employed in linked lists. Given the one-way nature of hash functions, where it is probabilistically infeasible to guess the function input given its output, the blockchain links based on block hashes guarantee that blocks already in the blockchain cannot be altered without detection. The hash of the content of each transaction is signed using asymmetric encryption and the participants are known using their PK^+. Figure 3.2 depicts a high-level view of the structure of the blockchain.

Each block consists of two main parts which are: transactions and block header. Recall from Section 3.1 that a transaction is the basic communication primitive between participating nodes in blockchain. Each transaction contains the following core information: *ID*, which is the transaction identifier and is the hash of the transition content; *Content*, which is the transaction content that depends on the application; *PK*$^+$, which is the public key of the transaction generator that essentially serves as the node's identity; and *Sign*, which is the signature of the transaction generator that is the signed version of *ID* by the PK^- of the transaction generator. A transaction may also contain other fields depending on the application; for example, a Bitcoin transaction contains an input field that is a link to the output of a previous transaction and spends the output in the current transaction.

The block header maintains the basic information about the block including timestamp, which denotes the time when the block was generated; Merkle tree root, which is the root hash of the Merkle tree built by the transactions (see the discussion below); *B_ID*, which denotes the block identifier which is essentially the hash of the content of this block; and *P_B_ID*, which is the identity of the previous block in the blockchain. Including the previous block hash in the current block effectively creates a tamper-proof chain between the blocks in the blockchain as the hash of each block contains the hash of the previous block in the ledger. Other header fields depend on the type of the blockchain and vary based on the application.

Recall that each block contains a Merkle tree that is essentially a summary of all transactions in the block and is constructed by repeatedly hashing the leaves in pairs until only one hash remains, which is known as the Merkle

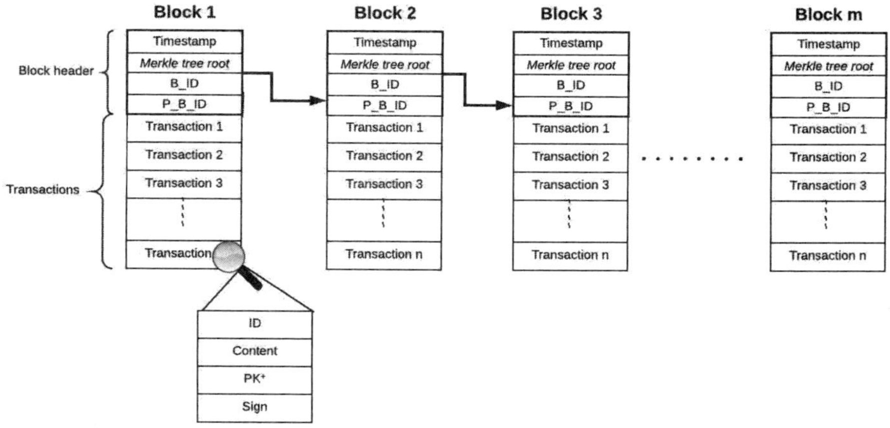

Figure 3.2 The structure of the blockchain.

tree root, as shown in Figure 3.3. The content of the transactions is stored in the leaves of the tree. The Merkle tree speeds up the process of verifying the existence of a transaction in a block. As a representative example, consider the Merkle tree shown in Figure 3.3. To prove that transaction D exists in the blockchain, the transaction generator must reveal D, h(C), and h(A|B). The verifier first calculates the h(C|D) and then the root hash of the Merkle tree and compares that with the root in hash stored in the block. If matched, the existence of the transaction in the block is proved.

Based on the visibility and access given to the users, blockchains can be categorized in two main groups: permissioned and permissionless blockchains. Permissioned blockchains only allow authorized nodes to join and participate in the blockchain. Additionally, the participating nodes may have different read-or-write access rights to the blockchain, which enables the blockchain managers to control the nodes that can mine new blocks. Hyperledger [5] is an example of a permissioned blockchain. A permissionless blockchain is a public blockchain where anyone can join the network and all participations have equal access that enables any node to participate in storing data in the blockchain. In general, permissioned blockchains are faster, have higher throughput, and incur less overhead as compared with permissionless blockchains given that there is a level of trust in permissioned blockchains to the authorities managing the chain. However, a permissionless blockchain assumes no trust between the participating nodes and enables untrusted participants to agree (achieve consensus) on the current state of the blockchain. This highlights an inherent coupling between the degree of trust among participating nodes and the amount of computational overhead involved in maintaining the blockchain, which we will revisit throughout the book.

3.2.2 Storing New Blocks

In this section, we outline the process of storing data in the blockchain. Figure 3.4 outlines multiple steps involved in the process of storing a new transaction

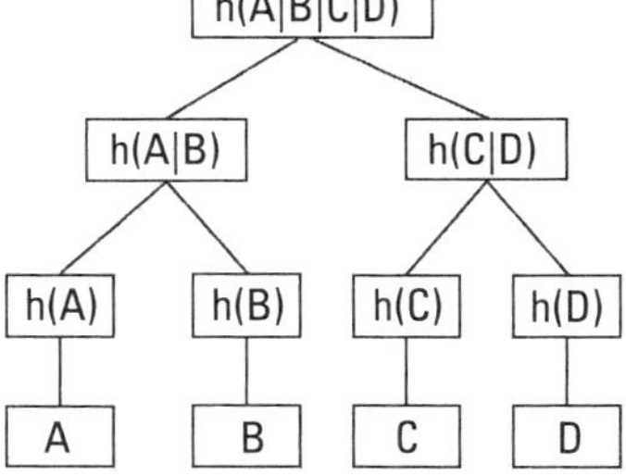

Figure 3.3 An example of a Merkle tree.

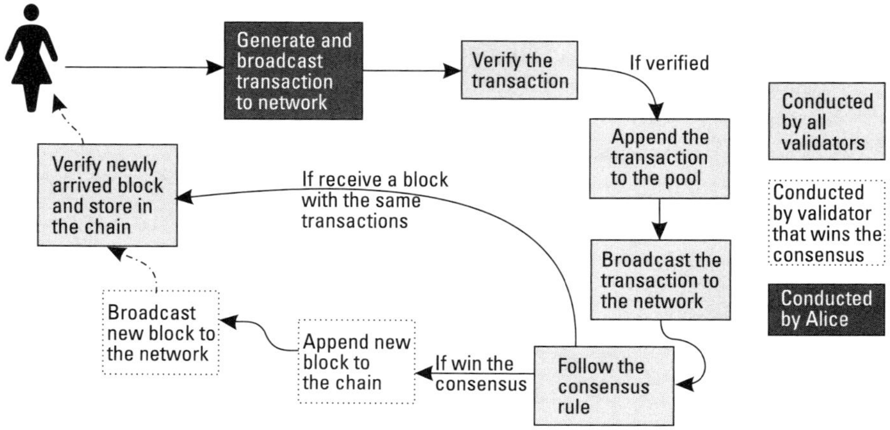

Figure 3.4 An overview of different steps involved in storing transactions in the blockchain.

in the blockchain. Assume that Alice is a member of a blockchain (either permissioned or permissionless) and generates a transaction to exchange information with other users. The transaction is signed using asymmetric encryption and is then broadcast to the blockchain. All participating nodes verify the transaction by checking Alice's signature with her PK^+ (which is included in the transaction). Other verification steps vary based on the type of the blockchain. The validators store pending transactions (i.e., transactions that are not yet stored in the blockchain) in a pool of pending transactions. When the size of the pool reaches a predefined value, known as *block_size*, the validator forms a block (see Section 3.2.1), which contains Alice's transaction, and permanently stores the block in the main chain by following the consensus rules. The consensus algorithm ensures blockchain security by introducing randomness among the validators, so that individual validators randomly add new blocks over time. The consensus algorithms also require the validators to choose a leader that can store the new block using a leader election algorithm, or provide a proof of X to prove that they have invested resources for storing the new block. Bitcoin, for instance, employs Proof of Work (POW) as the underlying consensus algorithm where the participating nodes must solve a hard-to-solve and easy-to-verify cryptographic puzzle before storing a new block in the blockchain. We provide a comprehensive discussion on existing consensus algorithms in Section 3.2.3.

New blocks are broadcast to the network and all participating nodes verify the block, involving two main steps: (1) the participants verify if the validator followed the consensus algorithm based on the relevant consensus rules; and (2) the participants verify all transactions in the block. Once the transaction is stored in the blockchain, Alice, or any other participant involved in the transaction, can accept the transaction.

Due to the distributed nature of the blockchain, multiple validators may simultaneously attempt to append the same block to the blockchain, which creates a fork in the ledger. As an example, in Figure 3.5, two separate blocks are chained to the first block, creating a temporary fork in the chain. In blockchain, all the participants must agree on a single chain. To address the forking challenge, the participating nodes rely on the longest ledger rule where the ledger with the larger number of chained blocks is accepted as the main chain. As an example, in Figure 3.5, the length of the gray blocks is smaller than the length of the main chain (white ledger) and thus the gray blocks are in a fork. The transactions in the forked blocks are normally considered as invalid transactions, although this may vary for some blockchain instantiations. To ensure that her transaction is not in the forked blocks, Alice waits for transaction confirmation, which typically involves ensuring that a certain number of blocks are chained to the block. Transaction confirmation rules are application specific. For instance, in Bitcoin, three blocks must be added after the relevant block before its constituent transactions are confirmed.

3.2.3 Consensus Algorithms

The consensus algorithm is key to the security of the blockchain as it ensures randomness among the validators and protects against malicious validators that may store fake transactions in the blockchain. The randomness introduced by the consensus algorithm protects against malicious validators that may continuously generate blocks to either receive block incentive or store fake transactions.

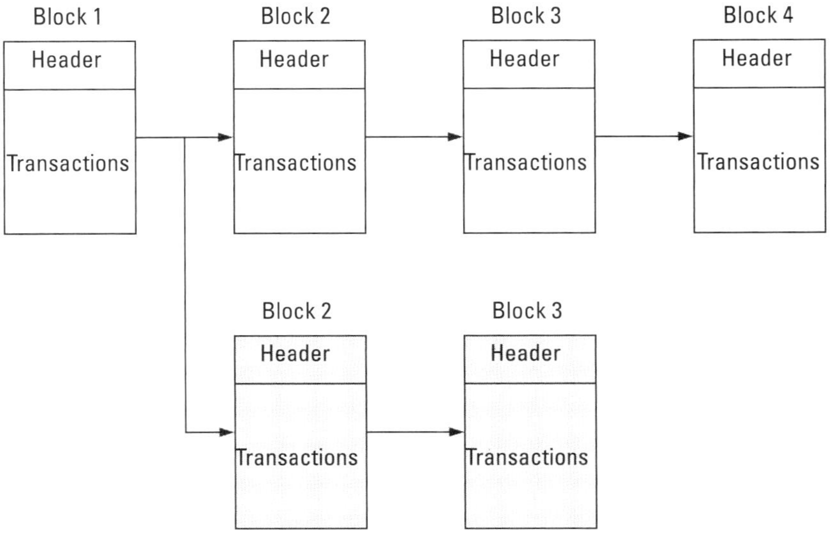

Figure 3.5 Forking in the blockchain.

In this section, we provide a review of the existing consensus algorithms in the literature, which are summarized in Table 3.1.

Bitcoin [1] employs POW as the underlying consensus algorithm. POW requires the validators to spend computational resources to solve a hard-to-solve and easy-to-verify puzzle. The validator that solves the puzzle first stores its block in the chain. To solve the POW, the validator tries to find a nonce value in a way that the hash of the nonce and the block content, that is, $H(Transactions \| Block_{header} \| nonce)$ starts with a particular number of leading zeros, known as difficulty. The difficulty of the POW is adjusted in 2,016 block intervals by the validators to ensure that only one block is stored each 10 minutes. This ensures that the puzzle is always difficult to solve and thus protects the security of the blockchain against the malicious validators that may attempt to double spend a coin. In a double-spending attack, the malicious validator attempts to store two transactions that use the output of the same transaction as input (i.e., spending a single coin twice). To do this, the malicious validator generates a block that contains the double-spent transactions and attempts to make it (or add it to) the longest ledger.

Table 3.1
Widely Used Consensus Algorithms and Their Features for CPS

Consensus Algorithm	Access	Control	Advantages for CPS	Disadvantages for CPS	Proportion of Honest Validators
POW	Public	Permissionless	—	Computational overheads, limited scalability, low throughput	>50%
POS	Public	Permissionless	Low computational overhead	Limited scalability, low throughput	>50%
FBA	Private/ consortium	Permissioned	Low computational overhead	High packet overhead, the miners must know each other, medium delay	3F+1
PBFT	Private/ consortium	Permissionless	Low computational overhead	High packet overhead, medium delay	3F+1
POA	Private/ consortium	Permissioned	Low computational overhead, medium packet overhead	The miners must know each other, medium delay	>50%
POET	Private/ consortium	Permissioned	Time-based lightweight	Hardware-specific	2F+1

The validator includes the final nonce value in the corresponding block header and broadcasts it to the network. The receiving validators verify that the block generator solved the puzzle by verifying that $H(Transactions \,||\, Block_{header} \,||\, nonce)$ starts with particular number of zeros that match the current difficulty value that is agreed upon by all the validators.

POW demands significant computational resources and limits the blockchain throughput (i.e., the total number of transactions that can be stored in the blockchain per second) to 7 transactions per second. Alternative consensus algorithms have been proposed to address the outlined limitations and to increase the blockchain applicability for real-world scenarios. In 2014, Ethereum [6] was proposed as a new blockchain instantiation that supports distributed applications by enabling the users to store and execute code in the blockchain. Ethereum uses Proof of Stake (POS) as the underlying consensus algorithm that demands validators to lock a specific amount of assets in the blockchain to be able to participate in storing a block. Unlike POW where the chance of winning the consensus depends on the computational power of the validators, in POS, the amount of assets each validator locks in the blockchain impacts the chance of winning the consensus and thus storing a new block. Given that the validators hold assets, they perform honestly in the chain so that they can protect their own assets and their corresponding value.

The authors in [7] proposed a Federated Byzantine Algorithm (FBA) based on the byzantine fault tolerance algorithm, which chooses the validator of the next block based on leader election algorithms, such as the one in [8]. The network validators in FBA are known to each other, which makes FBA suitable for permissioned blockchains. Each validator forms a quorum by selecting a random set of validators in the network. The validators then share their own quorum with other validators. The validators explore the intersection of the quorums and the node that is chosen by the largest number of validators is selected as the validator of the next block. In case no intersection is detected, multiple validators are selected, which leads to forking.

Hyperledger Fabric [5] is an open-source blockchain instantiation developed by IBM and employs the Practical Byzantine Fault Tolerant (PBFT) algorithm as the underlying consensus algorithm. In PBFT, all the participating nodes in the blockchain are ordered sequentially based on their timestamp and the node in the first place is selected as the primary node while others are called backup nodes. The primary node is responsible for storing new block to the blockchain and is changed during each period known as *view*. The users send their transactions to the primary node. The latter then sends the transaction to the backup nodes to be verified. The primary node waits until at least $3f + 1$ backup nodes return the same state about the transaction, where f is the total number of faulty nodes that the system can tolerate. In case a primary node refuses service to the blockchain users, the backup nodes change the primary node.

In [9], the authors proposed the Proof of Authority (POA) consensus algorithm, which is based on the Byzantine Fault Tolerance algorithm. POA reduces the messages exchanged between the validators by allocating weights to the identity of the validators (i.e., their reputation in the system). Thus, the identity of the validators must be known to all participating nodes.

Intel proposed a hardware-specific consensus algorithm for permissioned blockchains known as Proof of Elapsed Time (POET) [10]. In POET, each validator sleeps for a selected random time period. The validator that wakes up before other validators stores the block in the blockchain. A malicious node can compromise the security of the blockchain by claiming to have a short random time and thus storing new blocks ahead of other validators. To protect against this attack and to ensure that the random number is genuine, POET relies on the Trusted Execution Environment (TEE) in Intel CPUs. TEE allows the secure execution of code in a trusted tamper-resistant environment. TEE allows the validators to verify that a particular random time was the output of a code run in the TEE, which enables the participating nodes to trust on the random time claimed by the validators. Because it is dependent on hardware-specific features, POET is limited in its broad applicability to other CPS hardware platforms.

3.3 A Review on the Existing Blockchain-Based Frameworks for CPS

In this section, we briefly discuss four blockchain instantiations that are widely applied in CPS applications: Ethereum, Hyperledger, IoTA, and Corda.

3.3.1 Ethereum

Ethereum [6] enables the users to run code in the blockchain by introducing the concept of smart contracts. The smart contract contains the user code and is executable by all participating nodes in the blockchain. Recall from Section 3.1 that blockchain is immutable and, thus, altering the content is not possible. Thus, once a smart contract is stored in the blockchain, all participants can trust the contract and be assured that the contract output is not tampered with. Smart contracts are employed by the users to build Distributed Applications (DApp), which support a broad range of applications for blockchain. As an example, the list of authorized users to access specific assets can be stored in a contract and the contract can automatically authorize users to access specific assets. The user(s) can interact with the smart contract (e.g., send input or call functions) by sending transactions to the address of the contract, which is technically the hash of the contract output.

In recent years, multiple blockchain instantiations have been developed on top of Ethereum. Quorum [11] is a blockchain framework built on top of Ethereum that facilitates Ethereum adaptation in businesses. Quorum enables parties to encrypt data and share only with relevant parties though access control policies. Depending on the application, various consensus algorithms can be plugged into Quorum.

3.3.2 Hyperledger

Hyperledger [12] is a project founded by the Linux Foundation that supports the development of multiple blockchain-based distributed systems with the aim of improving the performance, scalability, and user privacy in conventional blockchains. Multiple companies joined the Hyperledger project, including IBM and Intel. Under the Hyperledger project, multiple application-specific blockchains have been developed. Hyperledger aims to provide tools, protocols, standards, and software for developers or companies to faster the development of blockchain-based architectures. Some of the frameworks developed under Hyperledger projects are the following. The Hyperledger fabric [13] was developed by IBM with support for plug-in consensus algorithms and various membership rules that give the system designers the flexibility to optimize the system based on their requirements. Fabric also supports the development of smart contracts. Hyperledger Iroha [14] was designed for infrastructural IoT and incorporates a new fault and crash tolerance consensus algorithm. Hyperledger Sawtooth [15] was originally contributed by Intel and incorporates the Intel consensus algorithm, known as POET (see Section 3.2.3). Sawtooth supports Ethereum smart contract and Solidity language through a transaction processing unit known as "seth." Sawtooth also supports multiple programming languages including Java, Go, Rust, and Python.

3.3.3 IoTA

IoTA is a new distributed ledger technology that eliminates the concept of blocks in conventional blockchains and relies on a Directed Acyclic Graph (DAG) to ensure scalability. A Tangle is proposed to achieve consensus among the participating nodes over the valid state and ledger of transactions. In IoTA, each node that generates a transaction must verify two randomly selected transactions so that its own transaction can be stored in the Tangle. This potentially removes the transaction fees and ensures high throughput for IoTA. We will study IoTA in detail in Section 4.2.6.

3.3.4 Corda

Corda [16] is a business-oriented blockchain solution. Corda enables businesses to run their own blockchain while introducing high interoperability, which enables transaction and asset exchange between multiple blockchains. Unlike conventional blockchains where the data related to smart contracts is broadcast in the network, Corda supports data on demand while enabling blockchain participants to verify the validity of transactions. Corda introduces state objects, which reflect the state of a smart contract. To increase scalability, in Corda, participants maintain the state objects in the place of a world-state in conventional blockchains.

In this section, we discussed four blockchain solution for CPS applications. Table 3.2 provides a high-level overview of the studied solutions. Next, we discuss the blockchain applicability through an example scenario.

3.4 An Example Scenario

In this section, we explain the fundamental advantages of applying blockchain in CPS by using supply chain as an example scenario. We provide detailed study on blockchain applications in a broader range of CPS applications in Chapters 8 to 11. Consider a food supply chain where a broad range of sensors capture various parameters, including temperature and humidity, in various stages of the supply chain. As outlined in Section 1.2.3, conventional CPS methods rely on centralized servers that collect and process the data. Blockchain eliminates the reliance on central trusted parties by enabling the participants to establish a trusted network even if they do not trust each other. The parties involved in the supply chain, which includes the sensor devices, farmers, and suppliers, are connected through the blockchain. Each participating node in the blockchain uses an anonymous pseudonym that is a public key as their identity, which enhances anonymity of the users. All communications between participating nodes and the data exchanges are recorded in the blockchain and are visible to all participants. To ensure privacy, the data associated with a device may be stored

Table 3.2
A High-Level Comparison of the Existing Blockchain Solutions for CPS

	Ethereum	Hyperledger	IoTA	Corda
Ledger Type	Permissionless	Permissioned	Permissionless	Permissioned
Token	Ether	None	None	None
Consensus	POW/POS	Pluggable	Tangle	Pluggable
Throughput	Medium	High	Self-scaling	Medium
Data Privacy	Yes	No	No	Yes

in a local storage, while the hash of the data is stored on chain, which ensures data integrity. Blockchain introduces high auditability and transparency, which enable the end users to trace and monitor various stages of the supply chain. Unlike conventional centralized methods, blockchain makes it impossible for any party, even SPs, to modify or remove data once it is stored on chain, which protects against malicious nodes that may attempt to alter data for their own benefit. A comprehensive study on blockchain applications in the supply chain is provided in Chapter 10.

Blockchain was first proposed as the underlying technology of Bitcoin. Most existing blockchain platforms have not been designed with CPS in mind, so there are many unresolved challenges for applying this technology in CPS contexts, which we discuss in detail in the next section.

3.5 Challenges in Adopting Blockchain in CPS

In the previous sections, we outlined the key features of the blockchain that make it an attractive solution to address security, anonymity, and decentralization challenges in CPS. However, applying blockchain in CPS is not straightforward and involves a number of challenges, which are outlined in this section.

3.5.1 Scalability

The pure distributed nature of the blockchain limits its scalability for large-scale networks, such as CPS, as the blockchain packet overhead increases quadratically with the number of nodes in the network. In conventional blockchain instantiations, all transactions and blocks are broadcast to all participating nodes, which consume significant bandwidth resources, while most of the CPS participants have limited bandwidth available. In CPS devices, the transmission units consume significantly more energy than processing units. This increases the energy consumption for broadcasting packets, while CPS devices have restricted energy supplies. Scalability improves the resources consumed by the CPS devices and thus ensures persistence (see Chapter 2).

3.5.2 Delay

Recall from Section 3.2.2 that storing a transaction in the blockchain involves following a consensus algorithm, which increases the delay associated with storing a transaction in the blockchain. Additionally, the users must wait for a confirmation time that protects the users against a double-spending attack. Recall from Chapter 1 that, in most CPS applications, the users and/or end systems expect real-time services, which are far beyond the latency of the existing blockchain instantiations.

3.5.3 Computational Resource Consumption

Blockchain removes the centralization by requiring all the participating nodes to participate in verifying all new transactions and blocks. In contrast, this potentially increases the processing overhead on the participants, particularly in large-scale networks with millions of transactions.

Recall that validators chose to store blocks in the blockchain after following a consensus algorithm. As discussed in Section 3.2.3, most of the existing consensus algorithms demand high computational resources. As an example, the hash rate to solve the Bitcoin POW is about 100,000,000 TH/s, which is increasing in time. However, this is far beyond the capabilities of CPS devices. POW is employed as the underlying consensus algorithm in 90% of the existing blockchain instantiations.

3.5.4 Memory Overhead

Recall from Section 3.1 that modification or removal of the previously stored data is impossible, which makes blockchain immutable and auditable. However, this potentially increases the size of the blockchain database in time (e.g., the Bitcoin blockchain database size is currently 270 GB) [17]. CPS contains millions of devices that correspondingly generate a significantly large number of transactions to be stored in the blockchain. Thus, the size of the blockchain in CPS is expected to be significantly larger than Bitcoin. To further exacerbate the memory overhead issue, most of the CPS devices have limited storage space available.

3.5.5 Throughput

Throughput is defined as the total number of transactions that can be stored in blockchain per second. The blockchain throughput is affected by the underlying consensus algorithm (e.g., Bitcoin throughput is 7 transactions per second). Generally, permisionless blockchains achieve lower throughput than their permissoned counterparts, given that the blocks in the latter are stored by trusted validators. However, with millions of nodes, CPS demands high throughput, which is far beyond the throughput in existing blockchain instantiations. New services and devices are introduced for CPS regularly, which frequently increases the throughput demand in such systems.

3.5.6 Privacy

The immutability of the blockchain makes it impossible for the users to remove the history of their transactions from the blockchain. While the data (if any) that is stored in transactions is encrypted, its persistence indefinitely creates long-term privacy risks for the data owners. Malicious nodes may attempt to deanonymize a user by analyzing the history of the transactions in the blockchain, which may compromise the user privacy. For example, in smart grid a malicious node may try to link multiple PK^+ to a single user, which reveals the energy consumption/generation pattern of the user and compromises the user's privacy. This may also lead to security and safety concerns (e.g., the attackers can identify the time that the house is vacant for robbery).

The immutability of the blockchain makes it challenging for the users to experience the right for their data to be forgotten and thus limits the blockchain compliant with the existing privacy regulations, for example, General Data Protection Regulation (GDPR) [18]. All enterprises that offer services to EU citizens must be compliant with the GDPR regulations, which raise questions about blockchain's broad adoption.

3.5.7 Reliance on Trusted Third Parties (TTPs)

Blockchain was first designed for cryptocurrencies where a digital asset is transferred between two users. Since then, it has been widely applied for trading other assets (e.g., cars or energy), which are represented as a digital asset in the blockchain. In these applications, the ownership of the asset is transferred through the blockchain and the physical asset is transferred off the chain. To ensure that both sides of a trade commit to their obligations, most of the existing works rely on a TTP (e.g., smart city manager or energy companies). However, reliance on a TTP raises challenges pertaining to centralization such as single point of failure and lack of privacy as discussed earlier. Note that in some CPS applications it may not possible to eliminate the TTP (e.g., energy companies must always be involved in demand and load data collection to ensure that energy supply is aligned with energy demand). However, it is critical to reduce the degree of reliance on TTPs to enhance user privacy and reduce the impact of centralization of TTPs.

3.6 Conclusions

This chapter has discussed blockchain for cyberphysical systems to conclude Part I of this book. It has defined the key elements of blockchain, including linked lists, encryption, and hash functions, and described widely used consensus algorithms and blockchain platforms in the context of CPS. While blockchain holds significant potential for addressing the challenges of CPS, discussed in Chapter 1, its adoption for CPS involves several challenges ranging from scalability to privacy and reliance on TTPs. The remainder of this book will explore solutions to these challenges that render blockchain more suitable for CPS. Part II will focus on general approaches that address the blockchain challenges for

CPS, covering the design of lightweight and scalable blockchains (Chapter 4), modifiable blockchains that promote storage efficiency and privacy (Chapter 5), trust in blockchain for CPS (Chapter 6), and anonymity in blockchain for CPS (Chapter 7). Part III of this book explores application-specific approaches for applying blockchain in CPS, covering smart grids (Chapter 8), smart vehicles (Chapter 9), supply chains (Chapter 10), and data marketplaces (Chapter 11).

References

[1] Nakamoto, S., "Bitcoin: A Peer-to-Peer Electronic Cash System," *Manubot,* 2019.

[2] Nguyen, T. V. H., et al., "Bitcoin Return: Impacts from the Introduction of New Altcoins," *Research in International Business and Finance,* Vol. 48, pp. 420-425.

[3] Koob, C., and D. P. Sonnier, "Method for Implementing Dual Link List Structure to Enable Fast Link-List Pointer Updates," U.S. Patent No. 7,111,289, September 19, 2006.

[4] Van Tilborg, H. C., and S. Jajodia, (eds.), *Encyclopedia of Cryptography and Security*, New York: Springer Science & Business Media, 2014.

[5] Cachin, C., "Architecture of the Hyperledger Blockchain Fabric," *Workshop on Distributed Cryptocurrencies and Consensus Ledgers,* Vol. 310, 2016.

[6] Wood, G., "Ethereum: A Secure Decentralised Generalised Transaction Ledger," Ethereum Project Yellow Paper 151, 2014, pp. 1–32.

[7] Mazieres, D., "The Stellar Consensus Protocol: A Federated Model for Internet-Level Consensus," Stellar Development Foundation, 2015.

[8] Sousa, J., A. Bessani, and M. Vukolic, "A Byzantine Fault-Tolerant Ordering Service for the Hyperledger Fabric Blockchain Platform," *2018 48th Annual IEEE/IFIP International Conference on Dependable Systems and Networks (DSN)*, 2018.

[9] De Angelis, S., et al., "PBFT vs Proof-of-Authority: Applying the CAP Theorem to Permissioned Blockchain," *Proceedings of the Second Italian Conference on Cyber Security,* Milan, Italy, February 2018, http://ceur-ws.org/Vol-2058/paper-06.pdf

[10] Chen, L., et al., "On Security Analysis of Proof-of-Elapsed-Time (POET)," *International Symposium on Stabilization, Safety, and Security of Distributed Systems,* 2017.

[11] Quorum blockchain, https://www.goquorum.com.

[12] Hyperledger project, https://www.hyperledger.org/.

[13] Hyperledger Fabric, https://www.hyperledger.org/projects/fabric.

[14] Hyperledger Iroha, https://www.hyperledger.org/projects/iroha.

[15] HyperledgerSawtooth,https://wiki.hyperledger.org/display/sawtooth/Hyperledger+Sawtooth.

[16] Brown, R. G., "The Corda Platform: An Introduction," 27 (2018): 2018, www.corda.net/content/corda-platform-whitepaper.pdf.

[17] https://www.blockchain.com/en/charts.

[18] Regulation (EU) 2016/679 of the European Parliament and of the Council 27 April 2016 on the Protection of Natural Persons with Regard to the Processing of Personal Data and on the Free Movement of Such Data, and Repealing Directive 95/46/EC (General Data Protection Regulation), EUR-Lex.

Part II

4

Lightweight Scalable Blockchains

4.1 Introduction

CPS comprises millions of connected everyday devices that sense the environment and share the sensed data with service providers (SPs), which process the data and offer personalized services to the end user. As outlined earlier in Chapters 1 and 2, existing CPS frameworks suffer from lack of security, privacy, high-resource consumption, and centralization. Chapter 3 outlined the salient features of blockchain as a potential solution to address the outlined challenges. We also discussed that applying blockchain in CPS is not straightforward and raises a number of challenges including low scalability, high packet and computational overhead, and high latency in storing transactions in the blockchain.

In this chapter, we first discuss existing lightweight blockchain solutions that attempt to address the outlined challenges in Section 4.2. We categorize these approaches into six categories: hierarchical approaches, optimized consensus algorithms, partial centralization, summarization, chain management, and new blockchain instantiations. Section 4.3 elaborates on a Lightweight Scalable Blockchain (LSB), an IoT-friendly blockchain instantiation that incorporates multiple optimization methods that include distributed trust, a distributed time-based consensus algorithm, and a distributed throughput management algorithm. Finally, Section 4.4 summarizes the key findings in the chapter.

4.2 Towards Lightweight Blockchain for CPS

In this section, we review the literature on various optimizations that aim to reduce the associated overheads of blockchain for large-scale networks including CPS. We categorize the existing works in six categories based on the optimization methods employed. We outline the key features of each optimization method and discuss recent works that employed these optimizations.

4.2.1 Hierarchical Approaches

In conventional blockchains, all transactions and blocks are broadcast and verified by all participating nodes. This, in turn, increases the packet and processing overhead and limits blockchain scalability. To address the outlined challenges, hierarchical approaches are widely applied in the literature where the participating nodes are clustered in different tiers and selected nodes manage the blockchain in each tier. Each tier is connected to the tiers above and below, which enables the participants to exchange information across multiple tiers. The grouping of the nodes may be based on various factors (e.g., resource availability, location of the nodes, or the access permissions to assets or data).

The authors in [6] proposed a hierarchical blockchain-based approach for access control management in IoT as shown in Figure 4.1. The proposed method constitutes three main layers. The first is the device layer, which is composed of all IoT devices and end users. The second is the fog layer, which includes devices with higher resource availability that manage transactions generated by a group of nodes in device layer. Each node in the fog layer, also known as the Security Access Manager (SAM), stores the transactions generated by the associated nodes in a local blockchain, which potentially reduces the delay

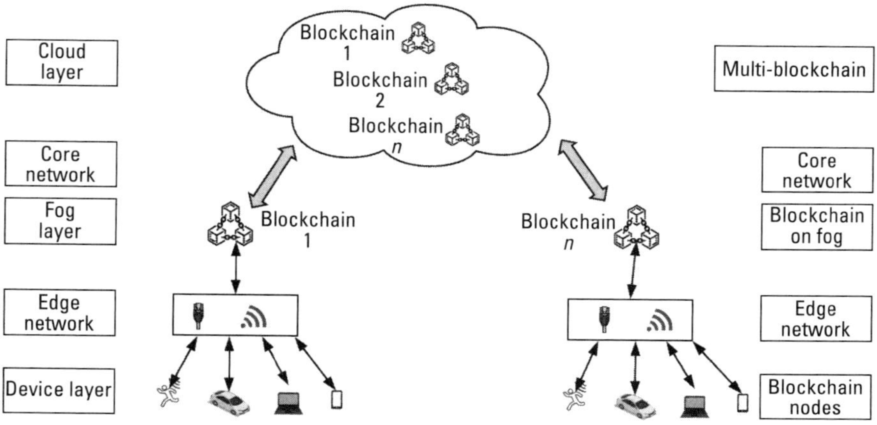

Figure 4.1 An overview of the proposed method in [6].

associated with storing transactions in the blockchain. The low-resource IoT devices use edge networks to connect to devices in the fog layer that perform the resource-consuming tasks including block verification. The third is the cloud layer, which comprises cloud servers with significant computational resources that are responsible for managing a global blockchain that stores the information of all the chains in the fog layer. SAMs in the fog layer periodically store the hash of their blockchain in the main chain to ensure integrity. The cloud servers use a core network to communicate with SAMs in the fog layer. The core network provides high-speed connectivity with high bandwidth that ensures high throughout and low latency for storing new transactions in the blockchain.

An IoT device can authorize any of the participating nodes to access the data collected by the device. To do so, the device populates the public key (PK^+) of the authorized node in a transaction and sends to its corresponding SAM. The latter verifies the signature of the IoT device and adds the transaction to its pool of pending transactions to be stored in the local blockchain. When a node (requester) wishes to access another node (requestee) in the network, the requester sends its request to the SAM responsible for managing transactions of the requestee. The SAM verifies if the requestee has previously authorized the requester to access the requestee. If so, permission is granted for the requester to access the requestee. In case the requester and requestee are deployed in different fogs, the grant permission transaction must also be signed by the servers in the cloud layer. The cloud server verifies the permission in the blockchain and signs the request transaction and pass the transaction to the SAMs. To store new blocks in the blockchain, SAMs attempt to solve a POW puzzle. To increase the privacy of the IoT devices, the devices are required to change their key stored in SAM periodically by sending an update transaction that contains the new $PK^+$of the same node.

The authors in [16] proposed a hierarchical blockchain framework for micro-insurance applications that offer pay-as-you-go insurance model for smart vehicles. The framework consists of three main tiers:

1. *Driver mobile app:* In this layer, a user-friendly mobile application is developed that facilitates the user interactions with the framework. The mobile application enables the driver to select a source and destination and departure or arrival times that are used by other layers to find a route and estimate insurance price.
2. *Micro-insurance analytic servers:* This layer analyzes the data of the users and calculates the best paths and the insurance price for each path. The layer encompasses three main components: route planning engine, pricing engine, and vehicle monitoring engine. The route planning engine processes the real-time traffic information, source and destination of the user, and departure or arrival time to calculate avail-

able routes toward the destination. The pricing engine uses the history of the accidents in each area, the user driving behavior, and the data provided by the IoT devices installed in the vehicle to estimate the price for each route. The data provided by the vehicle including the vehicle location and speed is monitored by the monitoring engine.

3. *Databases:* This layer includes the on-chain and off-chain databases that store the user data. Depending on its size, user data is stored in either of outlined databases. The hash of the off-chain database is stored in the blockchain to ensure integrity. The trip history and the premium payment made by the user are stored in the blockchain, which introduces high transparency. To manage the insurance, multiple smart contracts are deployed. For example, a smart contract is triggered by weather events that adjust the price of the insurance.

The three layers work collectively to deliver insurance services to the user while introducing high transparency in insurance industry. The authors implemented the proposed framework on top of Hyperledger Fabric and demonstrate the applicability of the framework.

In summary, the hierarchical optimization methods create multitier blockchains with limited number of nodes connecting to each blockchain, which reduces the processing and packet overhead. However, still the validators of the blockchains in each tier must follow the consensus algorithm before storing new blocks, which incurs significant overhead. To address this challenge, optimized consensus algorithms have been proposed. We discuss them next.

4.2.2 Optimized Consensus Algorithms

Consensus algorithms underpin blockchain security and ensure that untrusted participants can agree on a common state. POW, as the underlying consensus algorithm for over 90% of the existing blockchains, demands significant resources from the validators that are far beyond the capabilities of typical CPS nodes. Thus, multiple alternative consensus algorithms have been proposed to simplify the process of storing blocks and reducing the corresponding overheads. Note that permissioned blockchains can particularly benefit from simplified consensus algorithms given that there is a level of trust between the participating nodes. Recall that, in Section 3.2.3, we outlined multiple consensus algorithms with lower overheads as compared with Bitcoin POW including POS, POET, FBA, PBFT, and POA. Here, we focus on lightweight consensus algorithms that are specifically targeted at large-scale networks.

The authors in [8] proposed a green consensus algorithm known as Synergistic Multiple Proof (SMP) that aims to encourage the validators to collaborate toward solving the POW puzzle. SMP reduces the amount of wasted energy and resources in the validators, as compared with Bitcoin POW, by including the contribution each validator makes toward solving the POW to define the winning validator accordingly. The wining validator in SMP is selected based on a collaboration index (CI) and its computational power. There are two main approaches by which a validator can receive CI: (1) participate in solving the POW to store new blocks, or (2) chain a new block in the blockchain by winning the consensus algorithm, which is similar to the transaction fee in conventional blockchains. Similarly, the authors in [9] proposed Strongchain, an optimized version of the POW where the validators receive incentives for participating in solving the POW puzzle, even if they are not successful. The authors argued that POW suffers from lack of transparency as it is not possible for miners to estimate the mining processing requirement before storing new block as it completely depends on the block and hash function output and thus is random. To address this challenge, the authors defined a new hash target known as *weak target* that is smaller than the POW target (i.e., the strong target). When miners find a weak target, they add the target to the list of weak targets and broadcast the block to all nodes. By receiving the block, the miners verify the weak target using the same method used to verify the strong target. If verified, the miners add the weak target to the list of weak targets. Unlike conventional blockchains where the longest ledger is considered as the valid ledger, Strongchain defines weights for the ledgers and the ledger with the highest weight is considered as the valid ledger. The weight of a ledger is calculated by summing the total number of weak and strong targets in the ledger. This potentially makes the ledger that consumed the most computational resources as the valid chain.

In conventional blockchains, a malicious miner may perform selfish mining by keeping a parallel ledger of blocks and revealing the private ledger in an advantageous situation (e.g., when it will receive more block rewards). To protect against this attack and motivate the malicious miners to release their blocks, Strongchain awards the miners that solved weak targets independently in addition to the one that solved the strong target.

The authors in [10] proposed Bitcoin-NG, which amplifies the blockchain throughput and reduces the resource consumption by introducing two phases for solving the puzzle, namely, leader election and transaction serialization. During the leader election, a leader is selected to store transactions in the blockchain for a particular period of time known as an epoch. The leader serializes all transactions generated during the epoch and stores them in the form of microblocks in the blockchain. A microblock contains transactions and a block header that is chained to the previous block. The leader selection process can be offloaded to devices with higher computational resources (e.g., cloud servers to reduce delay and resource consumption). Given that a single leader is responsible for storing a new block during an epoch, the throughput is only impacted by the delay in communication and broadcasting blocks and transactions.

In this section, we studied the optimized consensus algorithms that reduce the processing overhead and delay associated with storing new blocks in the blockchain. The consensus algorithms ensure the blockchain security where untrusted participants are collaborating by demanding the validators to solve a puzzle and thus create a trusted network. To further reduce the processing overheads associated with the consensus algorithms, partial centralized blockchains are proposed that introduce a degree of trust between the participants.

4.2.3 Partial Centralization

Recall from Chapter 3 that the permissioned blockchains can afford the use of simplified consensus algorithms to reduce processing overhead and increase throughput. Permissioned blockchains shift from a purely distributed system to a decentralized system where selected nodes manage the blockchain, which reduces the packet overhead and increases scalability. In contrast, permissioned blockchains introduce a level of centralization and increase the privacy risks of the users as the validators may track the transactions generated by a particular user. The trade-off between the privacy and the blockchain overhead must be considered while employing permissioned blockchains.

The authors in [7] employed a permissioned blockchain as an access control framework for IoT devices. The proposed framework comprises six different participants:

1. *Wireless sensors:* These are low-resource IoT devices that sense the environment and provide data to SPs. Given that these devices have limited resources, they do not participate in managing the blockchain and are instead connected to the blockchain through managers.
2. *Manager:* This is a device that manages the access control policies for a set of IoT devices connected to it. During bootstrapping, the IoT devices must choose to join at least one manager in the network that ensures that they have connectivity to the blockchain. Similar to the sensors, the managers do not need to store the blockchain or participate in verifying transactions.
3. *Agent node:* This is responsible for deploying a smart contract in the blockchain.
4. *Smart contract:* This is employed by the managers to manage access control policies for the underlying sensors. The contract is deployed

during bootstrapping and the managers modify the policies by generating transactions.

5. *Blockchain network:* This is a permissioned blockchain where a set of authorized nodes manage the blockchain by storing new blocks and verifying new transactions and blocks.
6. *Management hubs:* These are resourceful devices that translate the packets generated by IoT devices into blockchain transactions and are directly connected to the validators. The proposed framework enables the IoT users to manage access to their devices while reducing the processing overhead on the devices for managing the blockchain.

The authors in [14] proposed a blockchain-based framework to share patient's data. The data of the patients is highly private, and, thus, the participating nodes in the blockchain are only authorized nodes by a legal authority (e.g., the ministry of health). The legal authority verifies the identity of the nodes that wish to join the blockchain. The data of the patients is stored off the chain to reduce the size of the blockchain and increase scalability. The proposed framework supports sharing of the patient's data with third parties (e.g., medical research centers). To do so, the third party sends a request to the hospitals to access particular type of data. The hospital then seeks further permissions from the users and shares data with the third party. The hash of the exchanged data is stored in the blockchain to ensure integrity.

In [15], the authors proposed a framework to exchange health data using private blockchain. The framework encompasses the users, doctors, wearable devices, and hospitals. The framework is implemented using Hyperledger Fabric. The data provided by the user or the wearable devices is stored in a cloud storage by a mobile application installed in the user device. Similar to other existing works, the hash of the exchanged data is stored in the blockchain to ensure integrity. The user can authorize other participants to access their data using an access control scheme that is implemented utilizing a membership function service in Hyperledger Fabric. The user defines the permissions associated with other participants, which are stored in the blockchain in the form of a transaction. The membership function service validates the access requests of the participants and grants permission accordingly.

In summary, the partial centralization methods introduce a level of trust between the participants and thus reduce the associated overheads in managing the blockchain. Storing all transactions in the blockchain in large-scale networks incurs significant storage, processing, packet overhead, and delay and limits the blockchain throughput. Thus, some related works consider storing part of the information (i.e., summary of the information or log of the data in the blockchain).

4.2.4 Summarization

In this optimization method, the participating nodes consolidate a group of transactions in a single transaction, referred to as the summary transaction, which reduces the memory footprint of the blockchain and increases the blockchain throughput as each transaction represents a group of transactions. Additionally, it enhances the user privacy as less information of the actual transactions is available in the blockchain. The authors in [2] proposed a blockchain-based auditing framework named Catena to increase the Bitcoin blockchain throughput and reduce the delay in storing transactions. An overview of Catena is shown in Figure 4.2. A log server initially stores a genesis transaction that has one unspent output that is used as the input of future transactions and thus establishes the chain. The rest of the transactions in the ledger have two outputs: one is the same as the output of the genesis transaction and the other is an unspendable *OP_Return* that is used to commit the log statement. Having a single unspendable output ensures that each transaction can only be chained to one other transaction and thus ensures a linear log of transactions.

The log server receives statements, equivalent to transactions in conventional blockchains, from the participating nodes, combines them as a single log file, and stores the log in the form of a transaction in the blockchain. The server forms a Merkle tree (see Section 3.2.1) of the received statements and signs the root of the tree. The tree is populated in a transaction and stored in the Bitcoin blockchain; thus, Catena potentially multiplies the blockchain throughput by the total number of leaves of the Merkle tree. Catena introduces a header relay network (HRN) that comprises a set of nodes that join the Bitcoin network as full nodes. However, HRNs do not attempt to solve the POW but just verify all new blocks. Once verified, the HRN broadcasts the header of the block to the Catena network, which is used to verify transactions. This reduces the

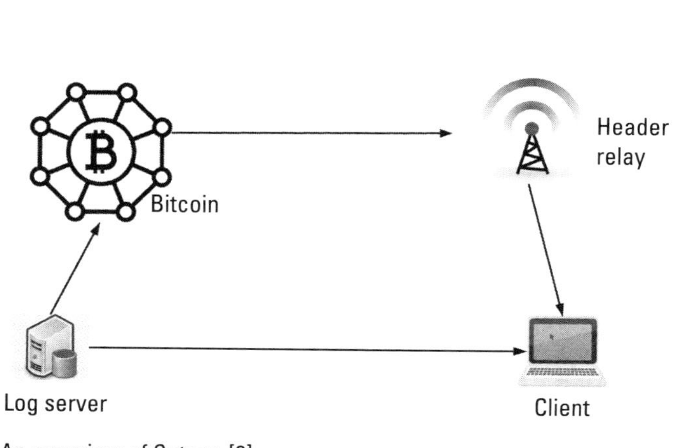

Figure 4.2 An overview of Catena [2].

blockchain memory footprint and packet overhead by nearly three orders of magnitude, from tens of gigabits to tens of megabits.

The process of auditing a previously stored statement is different in Catena as compared to conventional blockchains. In case a client needs to audit a statement, it requests to download and verify the block headers from the blockchain participants. Next, the client requests to download the statements corresponding to a transaction from the log server. The client verifies the statements by verifying the Merkle tree, that is, forming a new Merkle tree using the received statements from the log server and comparing the root with the existing root in the blockchain.

In a nutshell, the summarization optimization method reduces blockchain memory footprint and increases the throughput. However, it also reduces the auditability as limited information of each transaction is stored in the summary transaction. To reduce the processing and packet overhead while maintaining the auditability, chain management optimization methods are proposed in the literature.

4.2.5 Chain Management

Recall that the blocks are chained linearly and only transactions in the longest chain are accepted by blockchain nodes. To increase the blockchain scalability, new subset chains are introduced, namely, side chains and shards.

4.2.5.1 Side Chain

Side chains are separate chains that are attached to their parent blockchain and allow the users to transfer assets from the parent chain to and from the side chain in a particular rate through a two-way peg. The assets can be exchanged in the side chain independent of the parent chain, and, if needed, the assets can be returned back to the parent chain. Side chains are completely independent of the parent chain; thus, the side chains may employ different protocols (e.g., the consensus algorithm) compared to the parent chain. Additionally, a successful attack in the side chain does not impact the security of the parent chain. Side chains not only improve the blockchain throughput and scalability, but also increase the privacy of the users as the history of their interactions is no longer accessible by all participating nodes in the parent chain.

There are two main approaches to transfer assets from parent chain to the side chain, namely, centralized and distributed two-way peg. In the centralized approach, a trusted authority manages the asset transfer, where the user transfers the asset to the central authority. The trusted authority then transfers an equivalent amount of asset to the user account in the side chain. At the end, the user transfers the assets in the side chain to the trusted authority, which is then transferred to the user account in the parent chain by the trusted authority.

Multisig transactions are used to achieve distributed two-way peg. A multisig transaction requires n out of m ($n < m$) nodes to sign the transaction to be considered valid. It ensures that if any of the involved parties acts maliciously, the other involved parties can transfer the asset. In this method, a group of nodes are selected to manage the asset transfer. The asset is transferred to the address (i.e., PK) of one of the nodes that manages the asset transfer. The latter verifies that the asset is received and transfers an equal amount of asset to the side chain. Similar steps are employed in the reverse order to transfer the assets back from the side chain to the parent chain.

According to Back et al., side chains typically have the following properties [13]:

- Once the asset is transferred to the side chain, the owner of the asset must be able to move the assets back to the parent chain.
- Malicious nodes must not be able to block the transfer of the assets between the side chain and the parent chain.
- Asset transfer must be atomic meaning that either all assets must be transferred between the chains or none of the assets.
- Security vulnerability shall not create security risk in other chains.
- Side chains must remain independent meaning that the validators of the side chain must not require any information from other chains to verify transactions.
- The users should not be required to monitor the asset transfer in other existing side chains.

Back et al. [13] also proposed a solution to transfer assets in cryptocurrencies by providing proof of possession. To transfer assets, the asset owner locks the output of a transaction in the parent chain for a particular period of time known as the confirmation period. Additionally, the owner must solve a light version of the POW puzzle that protects against a Sybil attack. This protects against malicious nodes that may attempt a double-spending attack by spending the coin in the parent chain and the side chains. The user then generates a transaction in the side chain referring to the output of the locked transaction in the parent chain. The user must then wait for a consent period during which the user cannot spend his or her asset in the side chain. This provides further security against a double-spending attack. The asset can only be transferred from the side chains to the parent chain.

Poon and Dryja [3] proposed Bitcoin Lightning, which allows Bitcoin users to transfer a specific amount of coin to a side chain. A Bitcoin user (say, Alice) pays a specific amount of coin (say, X coins) to a known address. The central authority then creates a side chain with equal coins and transfers the coins to Alice. Alice is allowed to spend up to X coins in the side chain. Side chains can use simpler consensus algorithms than POW that enable Alice to pay small amount of money to users, known as micropayments, without paying transaction fees. In contrast, in the conventional Bitcoin network, micropayments incur significant transaction fees for the end user and require significant time to be confirmed in the blockchain. Once the payments are concluded, the final amount in the account of each participant in the side chain is reported to the central authority. The central authority verifies that the total amount of coin equals X coins. If so, the central authority transfers the coins to the corresponding accounts of the users in Bitcoin.

4.2.5.2 Sharding

In blockchain, sharding refers to partitioning the network in different groups (i.e., shards) where each group is only responsible for managing the transactions generated in its own shard but store the transactions from all shards. This significantly reduces the overhead in the participating nodes and increases the scalability. The transactions and blocks in each shard are shared across all shards, which ensure a distributed nature of the sharing.

The authors in [5] proposed RapidChain, which employs the sharding concept to enhance the scalability of the blockchain. The participating nodes in the blockchain form multiple committees (i.e., shards) and choose a random leader responsible for running the consensus algorithm and adding new blocks. To ensure distributed management of the blockchain, the committee leaders also share blocks among each other. During bootstrapping, the participating nodes agree on a set of nodes that function as the committee members and manage their corresponding committees (i.e., shards) and are known as the committee references, C_R. The members of C_R verify the transactions in the committee and store in the committee ledger. A node in a committee may generate transactions that must be sent to another committee. In this case, the C_R routes the packet toward the destination committee to be stored in the relevant ledger. In addition to the ledger in each committee, the C_R must agree on the state of the blockchain by running a consensus algorithm. RapidChain uses a synchronized consensus where the committee members agree on a digest of a block. Given that the size of the packets is small and only a fraction of the participating nodes run the consensus algorithm, RapidChain significantly reduces the associated delay and overhead.

A C_R or group of C_R may decide to collaborate and act maliciously to gain more benefits. To protect against this attack, in RapidChain, the C_R are reconfigured after a particular period of time, known as epoch time. At the end of each epoch, the C_R generate a random value and broadcast it to all committee members. This ensures that the committee members change in an unpredictable way. Note that transactions in blockchain may have outputs that are used as inputs in future transactions; for example, in Bitcoin, a node spends the output of a previously stored transaction and thus uses the output of the previous transaction as the input of its new transaction. This requires the verifiers to access both transactions. In sharding, a transaction might be stored in different committees; thus, the verifiers are required to access the transactions in other shards. To address this challenge in RapidChain, before storing a transaction (say, T_x) in the ledger, the committee member of T_x (referred to as C_{Tx}) sends T_x to the committee member that stored the transaction used as the input of T_x (referred to as C_{Ti}). If C_{Ti} verifies the transaction and stores it in the ledger, it sends a confirmation to C_{Tx}. Upon receipt of the confirmation, C_{Tx} stores T_x in the ledger.

In summary, the chain management optimization method introduces parallel chains to reduce the management overhead of the blockchain. Side chains enable the participants to transfer assets up to the amount that is transferred from the main chain; however, they require the participants to create a side chain and also are limited to cryptocurrencies. Sharding limits the number of transactions and blocks for which the validators are responsible; however, the validators still must perform the resource-consuming consensus algorithms and store the full version of the blockchain. Thus, blockchains, in particular, for large-scale networks like CPS, demand new instantiations that are designed based on the requirements of large-scale networks, which are studied in the next section.

4.2.6 Toward New Blockchain Instantiations

Due to the broad applicability of blockchain for nonmonetary applications, new blockchain instantiations have emerged that apply fundamental changes to the conventional blockchains to improve their performance. In this section, we study two such blockchains: IoTA and LSB.

IoTA [11] is a new Distributed Ledger Technology (DLT) that aims to increase blockchain scalability and remove transaction fees. IoTA achieves this by replacing blockchain with Tangle, which is essentially a new distributed ledger that eliminates blocks and thus chains transactions by employing DAG as shown in Figure 4.3. The latter introduces a directed graph with no loop that connects data structures (i.e., transactions). By removing the concept of blocks, IoTA also removes the need for validator nodes and transaction fees, as this fee was paid to the validators as compensation for the resources spent for following the consensus algorithm. Instead, the participating nodes in IoTA collaboratively manage the Tangle and achieve consensus on acceptable transactions.

The participating nodes in the Tangle generate transactions. In order for a transaction to be stored in the Tangle, the transaction generator must chain

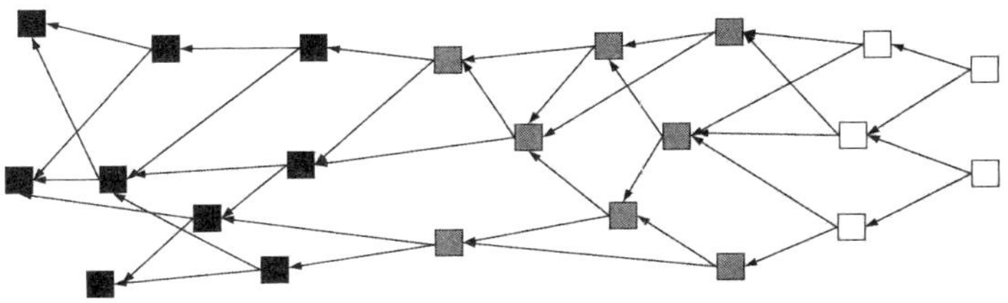

Figure 4.3 DAG employed in the Tangle.

this transaction with a minimum of two transactions. To chain a transaction, the transaction generator randomly verifies two other transactions from the list of unverified transactions (i.e., transactions that are not verified yet by any participant but are chained to other transactions) in the Tangle. The unverified transactions are the latest transactions that are stored in the Tangle; for example, in Figure 4.3 the white transactions are unverified. If verification is successful, the transaction is stored in the Tangle. Otherwise, the transaction generator selects another transaction from the list of unverified transactions to chain its transaction. As the number of participating nodes, and thus the number of transactions, in the Tangle increases, the number of verified transactions by the participants increases, which enables the Tangle to be self-scaling.

In conventional blockchains, the transaction recipient must wait for the confirmation time that protects against double-spending attack (see Chapter 3). Tangle introduces a weight for each transaction that is the total number of transactions that are chained (i.e., connected to the current transaction). Higher transaction weights indicate more trustworthy transactions. As an example, the white transactions in Figure 4.3 are not yet verified, while the black transactions have gained enough weight to be considered as valid transactions. The gray transactions are in between and gained a level of weight.

Recall that, in conventional blockchains, the validators must store the full history of the blocks to be able to verify new blocks, which consumes significant resources. Tangle does not require the participating nodes to store the full history of the transactions, but rather only a small part of the Tangle. This feature makes it possible for different participants to have different views of the Tangle. Thus, some nodes may not have particular transactions, or a branch of transactions, stored locally, which reduces the storage consumption of the Tangle. However, Tangle does not support smart contracts (see Chapter 3), which makes it impossible for the participants to run distributed applications.

The authors in [12] proposed a new blockchain instantiations optimized based on CPS requirements known as LSB. LSB is self-scaling, reduces packet and processing overhead compared to the conventional blockchains, and introduces distributed trust and consensus algorithms. We provide a detailed study on LSB below.

4.3 LSB for CPS

LSB is an IoT-friendly and self-scaling blockchain instantiation that maintains the core features of conventional blockchains including auditability, immutability, decentralization, and trust while significantly reducing the processing and packet overheads. In this section, we first provide a high-level overview of LSB in Section 4.3.1. We then discuss formation of the blockchain network in Section 4.3.2. The structure of LSB blockchain and the types of transactions are discussed in Section 4.3.3. Section 4.3.4 covers the process of storing a transaction in the blockchain, while Section 4.3.5 discusses the process for verifying newly formed blocks. Section 4.3.6 outlines load management in LSB, which underpins LSB's self-scaling feature. Finally, Section 4.3.7 discusses the transaction flow.

4.3.1 Overview

The CPS participants (e.g., sensors, devices, users, SPs, and cloud storages) jointly form an overlay network as shown in Figure 4.4. The participants manage a public blockchain by verifying and storing new transactions and blocks. Each node employs a changeable PK^+ as its identity, which introduces some level of anonymity. Recall from Chapter 3 that the fully distributed nature of conventional blockchains increases the associated packet and processing overheads and limits scalability. To address this limitation, in LSB, the overlay network is clustered, as shown in Figure 4.4, and only the cluster heads (CHs) participate in blockchain management. In CPS, the communication between devices might involve data (e.g., a camera sending real-time data to the user). In conventional blockchains, the data is stored within the transactions and broad-

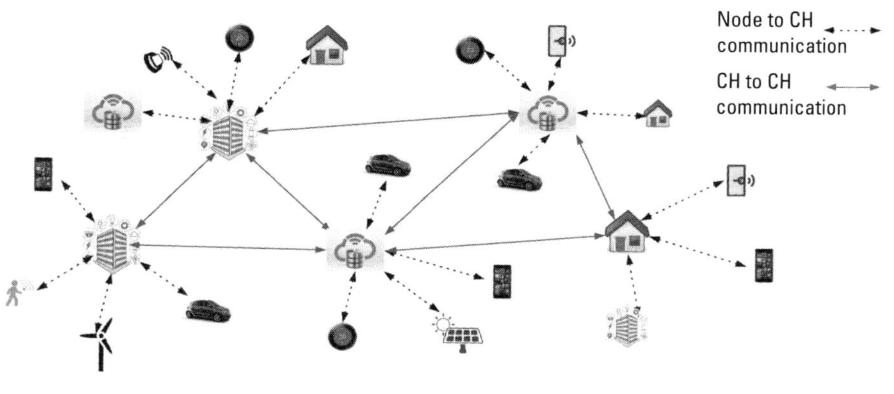

Figure 4.4 An overview of LSB.

cast to the network, which increases the blockchain packet overhead. To address this challenge, LSB separates the transaction and data flow in a way that the transactions are broadcast to all participations while the data is routed directly to the destination. The transaction contains the hash of the exchanged data to ensure data integrity.

LSB is designed for CPS applications and thus incorporates a number of optimizations. The first optimization is distributed trust. In conventional blockchains, all transactions in new blocks must be verified by the participating nodes before accepting the block, which ensures blockchain security against malicious validators that might store fake transactions in the blockchain. However, in large-scale networks such as CPS, the verification incurs significant processing overhead. To address this challenge, LSB introduces a distributed trust algorithm where the percentage of transactions that need to be verified in new blocks decreases as the CHs build up trust in each other. To ensure blockchain security, there must always be a particular number of honest CHs in the network as outlined later in Section 4.3.5. The second optimization is the distributed time-based consensus (DTC) algorithm. The DTC algorithm permits each CH to store one block in a specific time period, known as the consensus period. The consensus period is enforced by distributed neighbor monitoring where each CH verifies the blocks generated by its neighbor CHs. The third optimization is the distributed throughput management (DTM) algorithm. The DTM algorithm ensures that the load in the network does not deviate from the blockchain throughout. The CHs adjust throughput by tuning either the consensus period or the number of CHs in the network.

The first step in running LSB is for the overlay nodes to form clusters and select the CHs, which is outlined in detail below.

4.3.2 Overlay Formation

Conventional blockchains rely on a purely distributed communication model where all transactions and blocks are broadcast to all participating nodes. However, this model is unlikely to scale in CPS with millions of devices. LSB moves from a distributed model toward a decentralized model where the network participants form clusters and only CHs participate in broadcasting the communications. Figure 4.5 illustrates distributed vs decentralized network models.

The participating nodes employ a distributed clustering algorithm, an example of which is proposed in [1], to divide the network into clusters without relying on a central authority. Based on [1], each node populates the total number of its neighbors in a packet and broadcasts to its one-hop neighbors. Once a node receives the packets from all its neighbors, it selects the neighbor with maximum connectivity (i.e., with maximum number of neighbors), as the CH, which ensures that the CHs have high connectivity. In case two nodes have

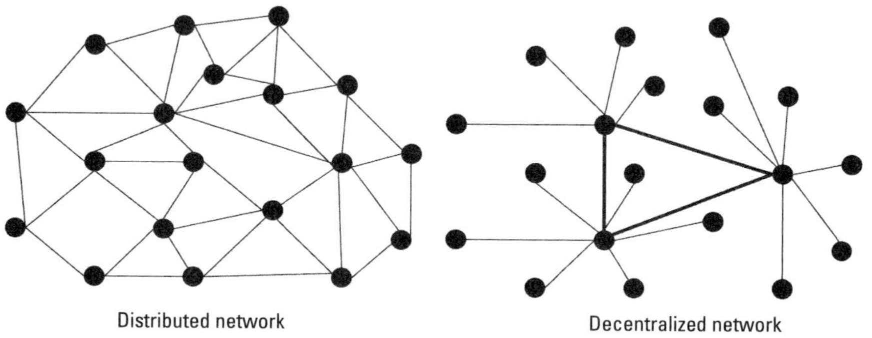

Figure 4.5 Distributed versus decentralized networks.

the same number of neighbors, the node connects to one randomly. In each cluster, the node with the second largest number of neighbors is selected as the co-leader. In case the CH cannot provide service to the cluster members (e.g., CH has lost connectivity), the co-leader functions as the CH to ensure that the cluster members always receive service.

To protect cluster members from receiving transactions from unauthorized nodes, LSB introduces an Access Control List (ACL) that is maintained by the CHs and contains PK^+ (of the cluster members of other nodes in the network) and the associated permissions. The cluster members can populate the ACL by their own PK^+ (say, $CM.PK^+$) that authorizes all transactions that contain $CM.PK^+$ to be sent to the cluster member, or by the PK^+ of any other node (say, $X.PK^+$) that authorizes node X to send transactions to the cluster member. The CH adds this information to the ACL and only sends transactions that match with the specified rules on the node. The ACL also protects against a denial of service (DOS) attack where malicious nodes flood a target node by sending a huge number of transactions so that the target cannot allocate resources to the genuine transactions. CHs do not send transactions to the cluster members unless a match is found in ACL. Additionally, each of the cluster members maintains a threshold for the total number of transactions received from the overlay in a particular period of time. If the number of transactions received from the overlay exceeds the threshold, the cluster member removes the keys in the ACL to block further transactions. The cluster member then populates the new keys in the ACL to authorize the permitted nodes to access its data.

Recall that, to reduce the packet overhead, LSB separates the transaction flow from the data flow in a way that the transactions are broadcast while the data is routed towards a particular destination. To route the data to the destination, the CHs employ conventional routing algorithms (e.g., OSPF [17]) to find a path toward another CH. The requester populates the ID of its CH in the transaction sent to the requestee that is used by the requestee to route data to the requester. To send the data packet back to the requester, the requestee populates the ID of the destination CH (i.e., the CH where the data requester is located) in the data packet and sends the packet to its own CH. The CH then uses routing algorithms to locate a path to the destination and then directly sends data to the destination CH, which then forwards the data to its cluster member.

4.3.3 Blockchain Structure

We now outline the basic structure of the blockchain and transactions in LSB. Recall that transactions are the basic communication primitives between blockchain participants. Blockchain users must first generate a genesis transaction, which essentially is the first transaction in a ledger. All transactions generated by the user are chained to the genesis transaction, which creates a history of transactions of the same user. A genesis transaction is considered as valid if it is coupled with one of the following:

- *Certificate of the burning coin:* In cryptocurrency, burning coin refers to paying a specific amount or fraction of coin to an unknown address that essentially permanently destroys the coin. The overlay node populates the address of the burn transaction in the genesis transaction and signs the latter with the PK^- corresponding to the PK^+ in the burn transaction, which serves as the certificate of the burning coin. To verify the certificate, the blockchain nodes must verify the existence of the burn transaction in the blockchain and match the signature in the genesis transaction with the PK^+ in burn transaction. If verified, the genesis transaction is stored in LSB.
- *Certificate from the certificate authority (CA):* The overlay node may also generate a genesis transaction by seeking a certificate from a CA. The CAs may employ various rules to issue a certificate. The PK^+ of the CAs are known by all participating nodes. To verify a certificate issued by a CA, the participating nodes match the signature in the certificate with the PK^+ of the CA.

The aforementioned steps for generating the genesis transaction limit the number of genesis transactions that an overlay node can generate and thus protects against a Sybil attack. In a Sybil attack, the malicious node pretends to be multiple nodes and floods the network with multiple transactions, which may lead to a DOS attack. The creation of multiple genesis transactions requires significant investment from the attacker. The fake transactions can be detected by the CHs during the verification of new transactions.

Based on the number of involved parties, transactions can be divided in two categories:

- *Singlesig:* This is a transaction which requires one party (i.e., the transaction generator) to sign the transaction to be considered as a valid transaction.
- *Multisig:* This is a transaction that requires *n* out of *m* participants to sign the transaction to be considered as a valid transaction. Multisig transactions are practical for cases where multiple parties are involved in a transaction. For example, if Alice wishes to buy goods from Bob, she can generate a multisig transaction that requires two out of three participants to sign the transaction. The third involved party can be local authorities that can intermediate in case of dispute.

The structure of a singlesig transaction is as follows:

$$T\_ID||P\_T\_ID||\text{Output}||\text{metadata}||\text{PK}^+||\text{Sign}$$

where *T_ID* is the transaction identifier that is essentially the hash of the transaction content. *P_T_ID* is the ID of the previous transaction in the same ledger of transactions as the current transaction (this can be the previous transaction generated by the same node). Output is the hash of the PK^+ that the transaction generator will use to generate next transaction in the ledger. This ensures that only the overlay node that has previously generated the corresponding genesis transaction is able to chain transactions in the ledger. Storing the hash of the PK^+ reduces the size of the transaction and protects against future attacks that may attempt to build PK^- using PK^+. The last two fields (i.e., PK^+ and Sign) refer to the PK^+ and signature generated by the transaction generator. The signature is essentially the encrypted hash of the transaction ID by PK^- of the transaction generator. The structure of the multisig transactions is the same as singlesig but with additional PK^+s and signatures of the involved parties.

In LSB, the cluster members directly send transactions to their corresponding CH. The CH first verifies the transaction (as discussed in Section 4.3.4). Next, the CH checks if the transaction is valid by checking the total number of signatures in the transactions and the number of signatures required to be considered as a valid transaction. If the transaction still needs to be signed by others, the CH looks for any match between the PK^+ in the transactions and the ACL. If a match is found, the CH sends the transaction to the corresponding cluster member to be signed. If there is no match or if the transaction is valid, the CH broadcasts the transaction to other CHs. The receiving CHs first verify the transaction and check if it is valid. The CHs add the transactions that are verified and are valid to a pool of pending transactions (i.e., transactions that are not yet stored in the blockchain). When the size of the pending transaction pool reaches a predefined size, known as *block_size*, the CH forms a new block. The structure of a block in LSB is as follows:

$$B\_ID \| P\_B\_ID \| TimeStamp \| B\_Generator \| Validators$$

where *B_ID* refers to the block identity which is the hash of the block content (i.e., all constituent transactions and block header fields). *P_B_ID* is the identity of the previous block in the blockchain. This essentially creates a chain between blocks that ensures blockchain immutability. Changing the content of a block (e.g., transactions) changes the corresponding hash, which will not match with the hash stored in the next block and is thus detected by the participating nodes. Timestamp is the time when the block was generated. *B_Generator* is the PK^+ of the generator of the block. The last field (i.e., validators) refers to the PK^+ of the CHs that have previously verified this block. We will further elaborate on this in Section 4.3.4. Once CHs populate the block, they start the process of storing the block in the blockchain.

4.3.4 Storing Blocks

Storing a block in the blockchain involves following a consensus algorithm that ensures the security of the blockchain. As outlined in Chapter 3, the conventional consensus algorithms demand significant resources and limit the blockchain throughput, which makes them inefficient for CPS. LSB introduces the DTC algorithm, which defines a consensus period during which each CH is allowed to store one block. Before storing a block, each CH checks if the following condition is met:

$$T_C - T_{P\_B} >= \text{Consensus period}$$

where T_C represents the current time and T_{P_B} is the last time when the same CH generated a block. This ensures that the CH generates one block per consensus period. The DTC algorithm introduces randomness among the CHs that attempt to store block simultaneously by defining a waiting period as:

$$0 <= \text{Waiting period} <= \text{Consensus period}$$

Before storing a block, the CH first randomly selects a waiting period and sleeps for that duration. At the end of the waiting period, the CH broadcasts the new block. A malicious CH may always claim a short waiting period to store its block ahead of other CHs. The honest CHs monitor the waiting period of other CHs. If a CH generates blocks with a short waiting period in multiple consensus period rounds, the honest CHs discard the malicious blocks and reduce the trust values for the malicious CH. It is possible that, during the waiting period, another CH, potentially with shorter waiting period, stores the same block. Once a CH receives a block that contains all or part of the transactions in its pool of pending transactions, it removes such transactions from its own pool of pending transactions as they are already stored in the blockchain.

A malicious CH, or a group of CHs, may attempt to generate more than one block in each consensus period. To protect against this attack, each CH stores the time stamp of the last block it received from other CHs that enables it to decide independently on accepting a new block. If the difference between the time stamps of the latest two blocks of a CH is less than the consensus period, the honest CH discards the block and reduces the trust value for the malicious CH. Once the CHs verify the consistency of the newly generated block, they verify the constituted transactions as outlined in next section.

4.3.5 Verifying Transactions

The CHs must verify all transactions in the newly created blocks to ensure that malicious CHs cannot store fake transactions in the blockchain. The verification of a transaction (say, *T*) involves the following steps:

- Match the signature of *T* with the corresponding PK^+.
- Verify that the *P_T_ID* of *T* exists in the blockchain.
- Verify that the output of the *P_T_ID* matches with the PK^+ of *T*.

Note that the verification of the transactions depends on the blockchain application and may involve additional steps. In conventional blockchain instantiations, all transactions in a new block must be verified, which increases the computational overhead in CHs (see Chapter 3). LSB uses a distributed trust algorithm where the percentage of the transactions to be verified in the newly stored blocks, and thus the processing overhead, gradually reduces as the CHs build up trust in each other. The distributed trust algorithm utilizes two metrics to identify the percentage of transactions that must be verified:

1. *Direct evidence:* CH A has direct evidence about CH B if it previously has verified blocks generated by CH B.
2. *Indirect evidence:* CH A has indirect evidence about CH B, if another CH has signed the current block generated by CH B as valid block. The CHs use indirect evidence only if they have no direct evidence about the block generator.

The CHs maintain a trust table, an example of which is shown in Figure 4.6. CH A records the total number of verified blocks generated by each of other CHs in the network. When the total number of verified blocks generated by CH B reaches a threshold, based on the values in the trust table, A verifies a smaller portion of transactions in subsequent blocks generated by B. If CH D generates blocks that contain fake transactions, other CHs reduce the trust value for D (i.e., total number of verified blocks).

The proposed distributed trust algorithm reduces the processing overhead. In contrast, it introduces security risks as only a fraction of transactions is verified, and thus a fake transaction may remain undetected by the honest CHs. The success rate of this attack depends on the values of the trust table (see the example in Figure 4.6). We studied the minimum percentage of transactions to be verified (PTV) based on the total number of CHs using NS3 simulator. We assumed that a malicious CH stores one fake transaction in a block and the block size equals 10 transactions. The simulation results are presented in Table 4.1. As evident from the results, the more CHs in the network, the smaller the value of PTV. However, a percentage of transactions in new blocks must always be verified even if the transaction is generated by a highly trusted CH. This protects against the on-off attack where a malicious CH alternates its behavior between being honest and malicious regularly to protect itself from being detected by other CHs.

4.3.6 Managing Load

As outlined in Chapter 3, the conventional blockchains suffer from low throughput (i.e., the limited number of transactions that can be stored in the blockchain). This potentially impacts the delay associated with storing transactions and thus increases the latency experienced by the end user (see Chapter 3). LSB introduces the DTM algorithm to manage the throughput. Each CH calculates a utilization parameter (α), which essentially is the amount of load generated in the network divided by the transactions that can be stored in the blockchain:

Direct evidence	**Number of previously validated blocks**	30	60	90	120	150
	Needs to validate	90%	70%	50%	25%	15%
Indirect evidence	**Percentage of the CHs that signed the block**	20%	30%	50%	80%	100%
	Needs to validate	90%	70%	60%	55%	50%

Figure 4.6 An example of the trust table maintained by CHs.

Table 4.1
The Minimum Percentage of Transactions to Be Verified Based on the Number of CHs

Number of CHs	3	5	10	15	20
Minimum PTV	80	60	40	20	10

$$\alpha = \frac{N * R * Consensus\_Period}{Block\_Size * CH\_Number} \tag{4.1}$$

where N denotes the total number of nodes in the network, R represents the average rate at which a node generates transactions, and *CH_Number* is the total number of CHs in the network. The main aim of the DTM algorithm is to ensure that α remains within a period of α_{min} and α_{max} that are defined based on the application. These values are within the range of [0, 1]. For example, if $\alpha_{max} = 1$, then the DTM algorithm adjusts the network throughput when the generated load in the network equals with the network throughput.

The DTM algorithm uses two parameters to tune α: *consensus_period* and *CH_Number*, which have indirect and direct impacts on the throughput. By decreasing the consensus period, the throughput increases, while, by decreasing the number of CHs, the throughput decreases. Each CH measures α at the end of each consensus period. If $\alpha_{min} > \alpha$ or $\alpha > \alpha_{max}$ for at least 2,016 consecutive blocks (which is inherited from the Bitcoin blockchain where the difficulty of the network is adjusted each 2016 blocks), then the CH attempts to measure new consensus period or *CH_number* in a way that $\alpha = (\alpha_{min} + \alpha_{max})/2$. Adjusting α to the average of the desired range reduces the chance of adjusting α again in the near future in case the load in the network experiences marginal changes. Changing the number of CHs requires restructuring of the network and thus incurs overhead. Thus, the CHs first attempt to adjust the consensus period. The consensus period must remain within the range of [2*Max_End-to-End_Delay, 10 min]. Maintaining the minimum value of the consensus period to two times the maximum end-to-end delay between CHs ensures that there is enough time for the newly generated blocks to propagate to the network. If the consensus period exceeds the desired range, the network is restructured.

When a CH receives a transaction from another CH that requests to adjust the throughput, it first measures α to check if it exceeds the thresholds. If so, the CH calculates α by using the suggested parameter values by the CH that generated the throughput adjustment transaction. If the recipient CH accepts the changes, it signs the transaction and broadcasts it; otherwise, the CH drops the transaction. Once the transaction is signed by more than 51% of the participating nodes, the changes are applied and broadcast to the network.

To evaluate the performance of the DTM algorithm, LSB is simulated in NS3 in a blockchain network with 50 nodes. The default value of the consensus period is 10 seconds. The simulation results are provided in Figure 4.7. Initially, the participating nodes generate 20 transactions per second that makes α above α_{max} and thus the consensus period is reduced. Later at $t = 21$ seconds, the number of requests again increases to 45 transactions per second that leads to further reduction of the consensus period to 2.5 seconds that is the minimum value of consensus period. The network load again increases at $t = 42$ seconds to 60 transactions per second. As the consensus period already hit the minimum value, the network is restructured, and new CHs are added to balance the load.

4.3.7 Transaction Flow

This section elaborates on the transaction flow in LSB in a smart vehicle ecosystem as shown in Figure 4.8.

The smart vehicles are equipped with a broad range of sensors that produce a large volume of data. The data may be stored either locally in the vehicle or in a cloud storage. The requester (e.g., the vehicle owner) is requesting data from the smart vehicle (e.g., the current location of the vehicle). It is assumed that the vehicle has previously authorized the requester to access its data by updating the ACL in its corresponding CH. The requester populates a multisig transaction that requires two signatures and sends to its own CH (step 1 in Figure 4.8). The CHs verify the transaction and, as it requires another

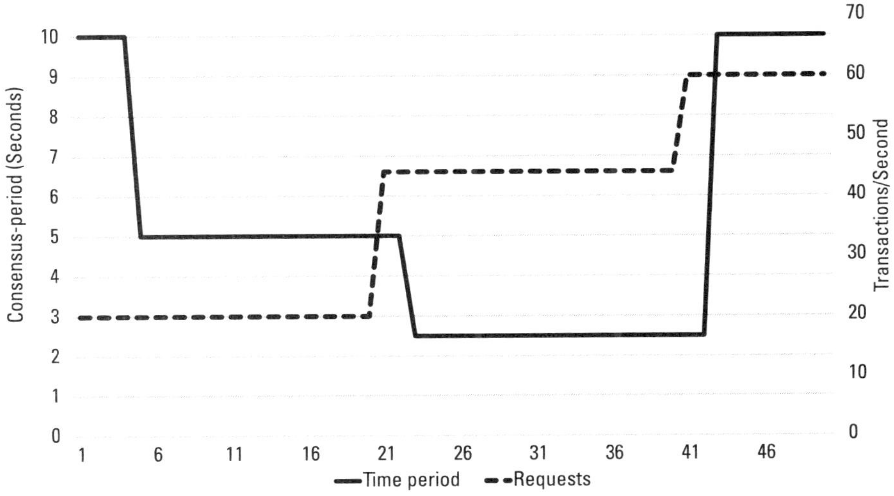

Figure 4.7 An evaluation of the distributed throughput algorithm.

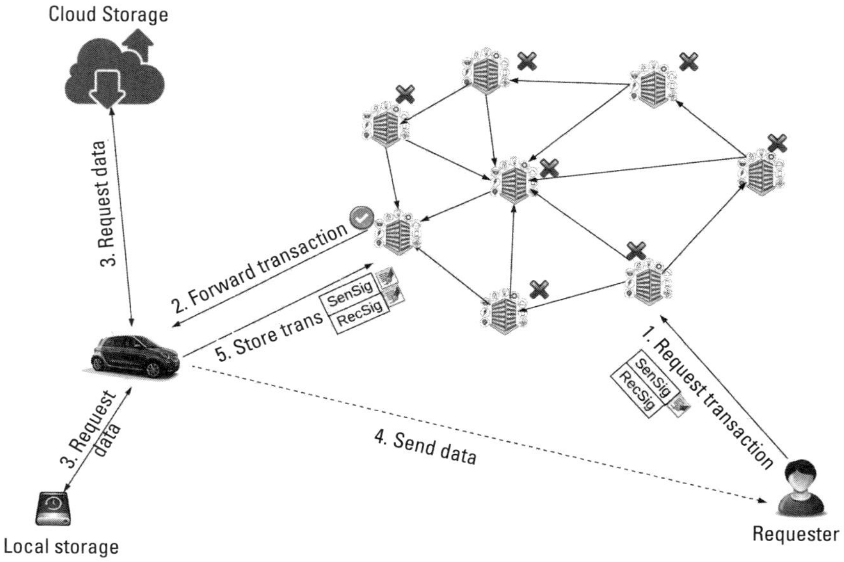

Figure 4.8 The flow of transactions in LSB.

signature, they broadcast the transaction until it reaches the CH responsible for the vehicle, which forwards the transaction to the vehicle (step 2). Depending on the data that is requested by the requester, the vehicle may need to fetch data either from a local or a cloud storage to share with the requester (step 3). The vehicle then directly routes the data to the requester (step 4). Finally, the requester signs the hash of the exchanged data in step 4 and populates that in the received transaction and send to its own CH to be stored in the blockchain (step 5). Given that both involved parties have signed the transaction, the CHs store the transaction in the blockchain. The stored transaction is employed as a reference of the exchanged data between the participants.

4.4 Comparative Evaluation

In this chapter, we studied the fundamental methods employed in the literature to reduce blockchain overhead and increase its scalability. We summarize the key features of each method in Table 4.2. We also study the compatibility of each of the optimization methods with the particular requirements of CPS and categorize their compatibility in three groups: high, low, and mid, where high has the highest compatibility with CPS and thus refers to the most suitable option.

Table 4.2
A Summary of the Optimization Methods Studied in This Chapter

		Compatibility with CPS Requirement		
Optimization Method	**Features**	**Throughput**	**Delay**	**Resource Consumption**
Hierarchical methods	Reduces the packet overhead by introducing multitier architecture; traffic in each tier is kept separate from other tiers; a separate blockchain may be used in each tier	Mid	Low	Low
Optimized consensus	Replaces the computationally demanding POW with more lightweight algorithms; reduces delay in mining new blocks and thus achieves higher throughput	Mid	Mid	High
Partial centralization	Introduces trusted authorities that participate in management of the blockchain; relaxes the security concerns in conventional blockchains and thus reduces the processing overhead and delay and increases throughput	High	Mid	Mid
Summarization	A single transaction in the blockchain represents a group of transactions, which increases blockchain throughput	High	Mid	High
Chain management	Introduces parallel chains to the main chain to record transactions between particular groups of the users	Mid	Mid	High
New blockchains: IoTA	Introduces the Tangle to eliminate mining process and fee by chaining transactions instead of blocks in conventional blockchains; achieves self-scaling feature and reduces the delay in storing transactions in the blockchain	High	High	High
New blockchains: LSB	Introduces hierarchical management of the blockchain to ensure scalability; introduces distributed trust to reduce processing overhead in storing new blocks; introduces time-based consensus algorithm to eliminate the computational resources demanded for mining; achieves self-scaling feature	High	High	High

4.5 Conclusion

In this chapter, we studied the existing lightweight solutions to increase the scalability of the blockchains. We categorized these methods in six main categories, namely: hierarchical, optimized consensus, chain management, partial centralization, summarization, and new blockchain instantiations. We provided a comprehensive study on LSB that addresses the challenges involved in the existing frameworks. In LSB, the CPS devices jointly form an overlay network where they manage a public blockchain. To reduce the associated overhead for managing the blockchain, LSB clusters the network and only the cluster heads manage the blockchain by verifying new transactions and blocks. LSB introduced a distributed trust algorithm that gradually reduces the processing overhead involved in verifying new blocks. LSB introduced a distributed time-based consensus algorithm that requires the cluster heads to wait for a particular time before storing a new block in the blockchain and thus significantly reduces the associated processing overhead. LSB also uses a distributed throughput management algorithm that ensures the blockchain throughput does not deviate from the network load.

In this chapter, we studied the existing solutions toward making blockchain scalable. As outlined in Chapter 3, adopting blockchain for CPS involves other challenges, including memory overhead and data persistence. In Chapter 5, we will study memory-optimized blockchains for CPS.

References

[1] Kousaridas, A., et al., "SYSTAS: Density-Based Algorithm for Clusters Discovery in Wireless Networks," *2015 IEEE 26th Annual International Symposium on Personal, Indoor, and Mobile Radio Communications (PIMRC)*, 2015.

[2] Tomescu, A., and S. Devadas, "Catena: Efficient Non-Equivocation Via Bitcoin," *2017 IEEE Symposium on Security and Privacy (SP)*, 2017.

[3] Poon, J., and T. Dryja, "The Bitcoin Lightning Network: Scalable Off-Chain Instant Payments," https://www.bitcoinlightning.com/wp-content/uploads/2018/03/lightning-network-paper.pdf, 2016.

[4] Jiang, Y., et al., "A Cross-Chain Solution to Integrating Multiple Blockchains for IoT Data Management," *Sensors*, Vol. 19.9, 2019, p. 2042.

[5] Zamani, M., M. Movahedi, and M. Raykova, "RapidChain: Scaling Blockchain Via Full Sharding," *Proceedings of the 2018 ACM SIGSAC Conference on Computer and Communications Security*, 2018.

[6] Ma, M., G. Shi, and F. Li, "Privacy-Oriented Blockchain-Based Distributed Key Management Architecture for Hierarchical Access Control in the IoT Scenario," *IEEE Access*, Vol. 7, 2019, pp. 34045–34059.

[7] Novo, O., "Blockchain Meets IoT: An Architecture for Scalable Access Management in IoT," *IEEE Internet of Things Journal*, Vol. 5, No. 2, April 2018, pp. 1184–1195.

[8] Liu, Y., et al., "LightChain: A Lightweight Blockchain System for Industrial Internet of Things," *IEEE Transactions on Industrial Informatics*, Vol. 5, No. 6, June 2019, pp. 3571–3581.

[9] Szalachowski, P., et al., "StrongChain: Transparent and Collaborative Proof-of-Work Consensus," *arXiv preprint arXiv:1905.09655*, 2019.

[10] Eyal, I., et al., "Bitcoin-NG: A Scalable Blockchain Protocol," *13th USENIX Symposium on Networked Systems Design and Implementation (NSDI),* Vol. 16, 2016, pp. 45–59.

[11] IoTA, https://www.iota.org/.

[12] Dorri, A., et al., "LSB: A Lightweight Scalable Blockchain for IoT Security and Anonymity," *Journal of Parallel and Distributed Computing,* Vol. 134, 2019, pp. 180–197.

[13] Back, A., et al., "Enabling Blockchain Innovations with Pegged Sidechains," http://kevinriggen.com/files/sidechains.pdf, *72,* 2014.

[14] Kim, K. J., and S. P. Hong, "A Trusted Sharing Model for Patient Records Based on Permissioned Blockchain," *J. Int. Comput. Serv,* Vol. 18, 2017, pp. 75–84.

[15] Liang, X., et al., "Integrating Blockchain for Data Sharing and Collaboration in Mobile Healthcare Applications," *2017 IEEE 28th Annual International Symposium on Personal, Indoor, and Mobile Radio Communications (PIMRC),* October 2017, pp. 1–5.

[16] Vo, H. T., et al., "Blockchain-Based Data Management and Analytics for Micro-Insurance Applications," *Proceedings of the 2017 ACM on Conference on Information and Knowledge Management,* November 2017, pp. 2539–2542.

[17] OSPF, https://tools.ietf.org/html/.

5

Memory-Optimized Blockchains

5.1 Introduction

Blockchain is an immutable database where the stored data can no longer be modified or removed from the main chain. Any modification of the block content (i.e., transactions) is detected by the participating nodes as the hash of the modified block will no longer match with the hash stored in the next block (in the *P_B_ID* field; see Chapter 3). Blockchain immutability introduces high auditability and security against malicious nodes that may attempt to alter the blockchain content. As outlined in Chapter 1, CPS comprises a million devices and applications that potentially generate millions of transactions. The CPS transactions may either contain data or be linked to data stored in a cloud by storing the hash of the data in the transaction. In both cases, the data must remain untouched to protect the blockchain consistency.

However, the immutability of the blockchain introduces a number of challenges for its adoption in large scale networks such as CPS. First, the size of the blockchain database will grow exponentially due to the huge volume of transactions being generated by the devices and users, which demands high bandwidth to transfer the blockchain and increases the cost of managing the blockchain. For instance, the Bitcoin blockchain size is about 250 GB [1], which is already substantial. In CPS the blockchain participants are expected to generate significantly larger volume of transactions and may outgrow the Bitcoin blockchain. Recall that the participating nodes in blockchain must store the blockchain to be able to verify new transactions. However, most CPS devices have restricted storage. Additionally, downloading such a large blockchain database demands significant bandwidth from the participants that is far beyond the available bandwidth of the CPS nodes that normally are equipped with wireless connection. Second, the nodes in CPS may generate transactions that should only be valid for a particular period of time (e.g., a user that purchased a smart thermostat may not need to share transactions with the service provider when the service period is finished). As another example, some devices may break down and thus the previous history of their transactions will no longer be needed. Furthermore, a user may install multiple devices in a facility that frequently generates transactions and the user may wish to summarize (i.e., consolidate) the transactions of all their devices into a single transaction. However, conventional blockchains do not offer such flexibility for the users to remove or compress their data in the blockchain. Third, the history of the user interactions is permanently stored in the public blockchain database. Although the user employs a unique identity (i.e., PK^+) to create each transaction, malicious nodes may attempt to deanonymize the user by linking multiple transactions in the blockchain, which potentially compromises the user privacy. This attack, known as linking attack, is often conducted against Bitcoin blockchain and leads to user deanonymization (see Chapter 7). Fourth, blockchain immutability makes it impossible to remove information and thus makes it challenging for the users to remove their data from the blockchain. Note that the right to be forgotten is one of the key requirements in the General Data Protection Regulation (GDPR) of the European Union (EU), discussed in [2]. All companies offering services to EU citizens must comply with this regulation, which imposes a hurdle in the broad applicability of blockchain. Similar legislation is also being drafted or enacted across other jurisdictions.

Having identified the challenges that emerge from blockchain immutability, the next section will discuss solutions that address the outlined challenges in isolation. In Section 5.3, we elaborate on Memory-Optimized and Flexible Blockchain (MOF-BC) that provides a comprehensive solution to flexibly managing blockchain content. Section 5.4 then summarizes the compatibility of the existing approaches with CPS requirements, while Section 5.5 concludes the chapter.

5.2 State-of-the-Art Memory-Optimized Solutions

This section provides an overview of existing approaches that address the challenges associated with blockchain immutability as outlined in Section 5.1. The state-of-the-art solutions can be categorized as off-chain data storage (Section 5.2.1), removing off-chain data (Section 5.2.2), data modification (Section 5.2.3), and optimizing blockchain transactions (Section 5.2.4).

5.2.1 Off-Chain Storage

One of the primary solutions proposed in the literature to reduce the blockchain storage size is to store the data corresponding to the transactions off-the-chain (i.e., in a cloud or local storage). To protect the data integrity, the hash of the data is stored in the blockchain.

The authors in [3] proposed a blockchain-based solution for data storage that is shown in Figure 5.1. The data of each user is distributed among a randomly selected group of participants while the hash of the data is stored in the blockchain to ensure integrity. The data is stored in a value-key based method utilizing distributed hash tables (DHT) that facilitate the data querying process. Each node in the DHT stores data corresponding to a specific key and is connected to the node that stores the values corresponding to the next key in the DHT. When a user query arrives, the DHT node checks if it possesses the data corresponding to the key value. If so, the node returns the data; otherwise, it forwards the request to the next node in the DHT. Only nodes authorized by the data owner are permitted to access a particular data source. To grant permission, the data owner generates a transaction authorizing a node (i.e., a PK^{+}) to access specific data.

The authors in [4] proposed a multilayer blockchain architecture where, similar to [3], the user data is stored by a selected group of nodes and the data access is managed through the blockchain. The proposed framework consists of a data plane and a management plane. The data of the users is stored and routed in the data plane. Similar to [3], the data is stored in a key-value form by all participants using the DHT. The key is the hash of <stream-ID, owner-ID, timestamp>. The user divides the data stream in different chunks each with a unique ID (i.e., stream-ID). Each chunk is then stored by a randomly selected group

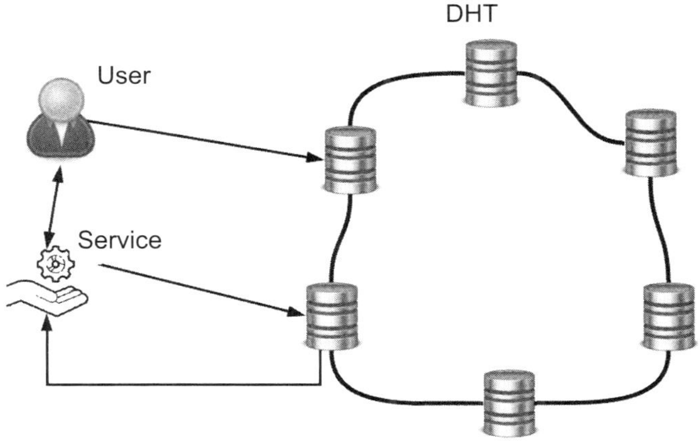

Figure 5.1 An overview of the proposed framework in [3].

of nodes. To retrieve data, the user must first locate a node that is storing the data by searching for the value tuple in the DHT. Then the user connects to this node to download the data, provided that the user has necessary permissions.

The management plane empowers the users to manage the access to their data using blockchain. The user authorizes a node to access specific data by generating a transaction that also contains the key to decrypt the data. To enhance the security of the framework against brute force attack to find the decryption key, a low-cost key-renewable method is proposed to renew the key in particular time periods. If the user wishes to revoke a previously granted access to a node, it encrypts the data with a fresh key, which prevents the nodes with the previous key from decrypting the data.

Theodouli et al. [5] proposed a framework for sharing personal data of patients. As shown in Figure 5.2, the framework consists of three layers. The first layer is the Web/cloud platforms that API used by the user to manage the data. The data of a patient is stored locally. The second layer is the cloud middleware that connects API in layer 1 to the blockchain in layer 3. This layer contains the Hypertext Transfer Protocol (HTTP) server that connects to the API in layer 1 and the smart contract API that connects to the blockchain layer. The third layer is the blockchain network that stores the blockchain. When the user produces data, the data is stored locally in layer 1, while the API stores the hash of the data in the form of a transaction in the blockchain. The patient grants permission to other users to access his or her data by generating permission contracts that are then stored in the blockchain.

Medical research centers may require access to a patient's data to conduct research. However, this may potentially compromise the privacy of the patient

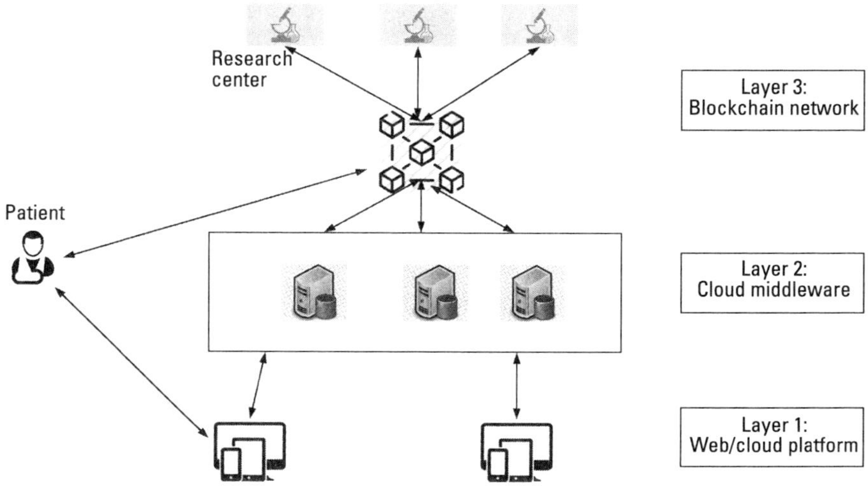

Figure 5.2 An overview of the proposed framework [5].

as the shared data is highly privacy sensitive. Recall that blockchain users are known by a changeable PK that potentially introduces a level of anonymity. In the solution presented in [5], the research centers can search the blockchain to find the PK of the users that own data. Next, the research center generates a smart contract and sets a "status" field as "pending" that shows the contract is not yet verified. If the user agrees to share data, he or she sets the status as "accepted." The smart contract also includes the permissions on how and for how long the requester can access data. The requester then connects to the layer 1 API to download the data. The hash of the exchanged data is signed by layer 1 API and the requester and is stored in the blockchain, which introduces auditability.

In addition to academic works, the need to optimize the blockchain database size has received attention from Bitcoin designers. Bitcoin 0.11.0 [6] was released with a built-in storage optimization method known as pruning. The pruning function allows the Bitcoin participants to remove the old blocks in the blockchain while still being able to verify the state of the blockchain. The nodes can choose the amount of storage space that they are willing to dedicate to the Bitcoin blockchain but must at least allow for 550 MB, which equals the size of 228 blocks. Storing the latest 228 blocks ensures that the nodes can verify the input and outputs of the newly generated transactions. Once the size of the blockchain database exceeds the defined value by the user, the wallet software removes the oldest blocks.

In summary, storing data off-the-chain while maintaining hashes on the chain has been one of the main approaches to reduce blockchain size. Another related approach has been to prioritize the storage of more recent transactions, while pruning away older transactions. The storage of data off-the-chain results in significant reductions to blockchain size, yet it does not address the privacy issues relating to data persistence on the blockchain, which we discuss next.

5.2.2 Removing Off-Chain Data

The GDPR is a law enforced by the EU to protect the privacy and user control over their data in the European Economic Area (EEA). Any company running services in EU or offering service to EU citizens must comply with GDPR. The right to be forgotten is one of the requirements of the GDPR that refers to users' ability to remove their personal data from the service provider servers. However, as outlined in Section 5.1, blockchain inherently does not satisfy this requirement due to its immutability feature.

To address this challenge, the authors in [7] proposed a data storage model that is applied on user health data. The participating nodes are divided into two groups: data controllers and data consumers, where a single node can be a member of both groups. To protect the privacy of the users, the personal data is stored locally by the data controller. The data controller maintains a list of the type of the stored data that is known as the local catalog. A copy of the local catalog is stored in the blockchain, known as the public catalog, that enables the users to search for a particular type of data and thus facilitates data querying. When the user generates new data, the data controller creates a new local catalog. The data controller then generates a transaction that maintains a key value:

$$key = (hash(encrypt(data\_key,\ data\_item\_id,\ bc\_catalogue\_id,\ catalogue\_id)))$$

where *data_key* is the hash of the key used to encrypt data, *data_item_id* is a unique identifier for the data, and the two other fields refer to the identity of the catalog that stores the type of data in the blockchain and locally. The transaction is stored in the blockchain as a reference to the data. To query data, the data requester issues a study transaction that contains the ID of the data type that it is requesting. This is then stored in the blockchain and the data controllers that store the requested data type reply to the data requester.

The proposed framework achieves the right to be forgotten for the users, and thus supports GDPR, by making the transactions stored in the blockchain independent of the data content stored by the data controller. The users can request the data controller to remove their data from the local storage. This potentially does not impact the transaction that is stored in the blockchain as the value is independent of the content of the transaction.

The authors in [8] proposed a privacy-preserving framework for smart city users also known as *privysharing*. Different application scenarios are defined, including smart grid, e-health, traffic management, and banking. To protect the user privacy, the communications and data exchanges between the participants in each application are limited to a private channel where only selected nodes can join. A membership service provider (MSP) manages the list of the authorized nodes to join the network with their corresponding permissions. The nodes in a channel have different read/write permissions that further increase the user privacy as data can only be read by authorized nodes. An access control list (ACL) is employed by the private blockchain where the users can manage accesses to their data.

Privysharing enables the users to monetize their data and thus introduces a new coin known as privycoin that is issues by the government authorities and service providers. To buy data from other users, the data requester must pay the data price through the blockchain to the data generator. Assume that Alice generates a transaction that contains data to be stored in the blockchain. Privysharing introduces two modes for storing the transactions: plain transaction flow and private data flow. In the plain transaction flow, Alice sends her transaction to a number of endorsers to verify based on the predetermined rules stored in a smart contract. If verified, the endorsers sign the transaction and send it back to Alice. Next, Alice sends the transaction to ordering nodes that form new blocks and store in the blockchain. Finally, the transaction is sent to committing peers to validate.

In the private data flow mode, Alice sends her data to an authorized subset of the endorsers that are predefined based on the network policies to protect the user privacy. The endorsers verify the transaction and store the corresponding data in a local temporary database. The signed transaction is then sent back to Alice with a reference to the data stored in the database that is technically the hash of the data. Alice then sends the transactions to the ordering nodes to be stored in the blockchain and then the committing nodes to be verified. The committing nodes can verify the transaction without knowing the corresponding data, which potentially protects the user data from being exposed to the committing members.

Different actions conducted on the data are recorded in the form of transactions in the blockchain that introduce auditability and enables the participants to monitor the access to their data, which enhances their privacy. The data of the users is stored in a separate database known as a world state. The removal of the user data in the world state does not impact the blockchain consistency that enables the privysharing users to experience the right for their data to be forgotten and thus makes it compatible with GDPR requirements.

Rieger et al. [9] studied the blockchain limitations with GDPR and proposed solutions to address such limitations. The authors highlighted that, to be compatible with GDPR, blockchain users must be able to remove their information and control who processes their personal data, the result of the data processing, and the periods during which the data is being accessed and/or stored by third parties. To address these limitations, the authors proposed three potential solutions:

1. *Central authority approach:* In this approach, a trusted central authority controls the flow of data and ensures that the data processes comply with GDPR. The authority establishes a smart contract that contains the agreements for data processing with each and every participant in the network, which forces the participants to follow the lawful rules identified in the smart contract and are enforced by law. In case the users demand to remove their data from the blockchain, the authority must erase the user data from the blockchain and recalculate the hashes in the blocks to ensure blockchain consistency.
2. *Shared responsibility approach:* This approach is similar to the central authority approach but, instead of a single node, a group of trusted nodes control the data flow and manage the smart contracts. By dividing the tasks between a group of nodes, this method aims to achieve higher scalability and reduce delay in processing user requests com-

pared with central authority approach. Additionally, this method addresses the issue of single point of failure that is inherent with the central authority approach. In this approach, trust is also decentralized, which increases the attack surface for malicious agents that may attempt to compromise the system.

3. *Pseudonymization approach:* In this approach, the data is stored anonymously using pseudonyms where the data owner can only be identified by knowing particular off-chain data. A group of trusted nodes known as controllers jointly manage the process of anonymizing data and access to the pseudonymized data. The users send their access requests to the controllers, who verify the request based on GDPR rules and grant permissions accordingly. If users request their data to be removed from the blockchain, the controllers remove the off-chain data about the users and thus potentially break the link between pseudonymized data and a specific user. This empowers the users to experience the right for the data to be forgotten.

Based on the outlined concepts, Germany's Federal Office for Migration and Refugees, known as BAMF (Bundesamt für Migration und Flüchtlinge), designed a GDPR-compliant blockchain-based framework. The BAMF framework consists of three main tiers: backend, integration, and blockchain. The backend tier contains the workflow and the user data, which are stored by the authorized nodes. The integration layer facilitates the communication and data exchange between the blockchain and the backend layer and consists of two subset layers: dashboard and privacy services. The dashboard facilitates the communications between the blockchain and the backend layer. The backend nodes submit the data events to the dashboard layer, which is then converted to the format specific to the blockchain and is stored in the chain. The blockchain users can also explore the data of the users in the backend through the dashboard web interface. The privacy services store a mapping between the pseudonyms in the blockchain and the real identity of the users in the backend. The blockchain layer is employed to distribute the pseudonyms and data events. The information stored in the blockchain does not contain any personal data that may compromise GDPR.

The users may demand to remove their information from the blockchain. To remove information, an authority issues a removal request to the privacy services. By receiving the request, the services remove the mapping information and store a remove event in the blockchain. The latter invalidates the pseudonymizers employed by a specific user and prevents any user from using these in their future transactions. The BAMF blockchain also introduces rectification procedure where the users can amend information stored in the blockchain by submitting a rectification event. The latter first removes a previous transaction from the blockchain (as outlined above) and then stores the new transaction.

Al-Zaben et al. [10] proposed a framework to manage personally identifiable information (PII) that is compliant with GDPR regulations. The blockchain framework consists of controllers, processors, and users. The controllers are legal authorities that issue PII for the users and processors are those entities that require processing PII and identifying users. The data of the users is classified as PII and non-PII where the former is stored off-the-chain and the latter is stored in the blockchain. Each user signs a smart contract with the controllers that outlines the authorities of the processors and type of process that can be conducted on the data. The controllers may request multiple PIIs of the users and share those with processors to process. The permission to access the data is granted once the controller request is authorized by the smart contract. The users can remove information from the blockchain by sending a request to controllers and processors. The controllers verify if the requested user is the owner of the data by matching the signature with the PK. If so, they remove the data stored in the local copy of the blockchain; however, the hash of the PII stored in the blockchain remains permanently in the chain.

In sum, we have seen that there are several methods for removing off-chain data to protect user privacy that provide data owners with more control over how and for how long their data is stored and controls access to their data. One issue regarding privacy is that multiple copies of the data may be maintained off-chain, which makes it difficult to verify if and when all copies of the data have been removed. Another consideration is that data owners may require modifying data on the blockchain for some applications. We discuss proposed approaches for data modification next.

5.2.3 Data Modification

As outlined earlier in Section 5.1, the immutability of the blockchain makes it impossible to modify previously stored data in the blockchain as the hash of the modified block will no longer match with the hash stored in the subsequent block, which compromises blockchain consistency. The authors in [11] proposed a solution to modify the blockchain content while maintaining its consistency. Instead of relying on conventional hash functions, the proposed method employs Chameleon hash functions to create the hash of a block. The Chameleon hash function was first proposed in [12] and has the same properties as the normal hash function except that it enables the participating nodes that know a trapdoor key to find a collision in the hash function. This empowers the users to create the same hash output using different input values. The key algorithms that are part of the Chameleon hash function include:

- *Key generation (KG):* This algorithm accepts a secret value (*k*) as input and outputs a hash key (*hk*) and a trapdoor key (*tk*). *hk* is used to hash a message, while *tk* is the trapdoor key that enables the key owner to modify the content of a transaction without changing the hash value.
- *Hashing:* This algorithm accepts an *hk* and a message (*m*) and generates a hash value (*h*) and a check string (*e*). The check string is used to verify the hash function and enables the nodes to modify content without breaking the consistency of the hash.
- *Verification:* This algorithm accepts *m, h,* and *e* as inputs and returns 1 if *h* is a valid hash of *m*. Otherwise, the algorithm returns 0.
- *Collision:* This algorithm finds a collision for a hash output. The algorithm accepts *tk, h, m, m`*, and *e* as inputs where *m`* is the new message to be hashed. The algorithm finds *e`* in a way that *verification(hk, m`, (h,e`)) = 1.*

As outlined above, only the user that knows *tk* can create a collision. Thus, by protecting access to *tk*, it is possible to prevent any modifications of transactions by malicious nodes. Recall from Chapter 3 that the validators store the hash of the block as a field in the block header. In the proposed method, the hash of the block is calculated using the Chameleon hash function. The authors studied block modification in centralized and distributed modes. In the centralized mode, a central controller in the network is responsible for modifying the blockchain content and thus knows the trapdoor key corresponding to the hash function used in the blockchain. To modify the blockchain content, the central authority accepts three inputs: (1) a chain of blocks in the blockchain to be modified, (2) the modified version of the chain, and (3) the trapdoor key that is stored locally by the central controller. The central controller uses the collision algorithm to find *e`*. The new ledger is broadcast by the central controller as a special ledger and all nodes in the network must accept and replace the old ledger with the new ledger. In the distributed mode, a group of nodes are selected initially during network bootstrapping. These nodes are trusted parties that communicate in a secure manner. Assuming that *n* trusted node participates in the network, the trapdoor is divided into *n* parts, each stored by one of the trusted nodes. To modify the content of the block, each trusted node performs the modification processes to find *e`* as outlined above. The nodes then broadcast the modified block signed by their share of the trapdoor key. The final block will contain the signature of all the trusted nodes and thus can be considered as valid by the network participants.

In a nutshell, data modification methods enable the users to modify the content of the blockchain while preserving the consistency of the chain. In contrast, this method limits the auditability of the blockchain as the users may change the content of their transactions without impacting the hash of the data that makes it impossible for the blockchain participants to track changes in the transactions. Another method is to optimize transactions by consolidating multiple transactions in one transaction that is studied next.

5.2.4 Optimizing Transactions

The optimization of transactions refers to combining multiple transactions to form one transaction that can be performed either before or after storing transactions in the blockchain. The optimization potentially reduces the size of the blockchain, as fewer transactions are stored, and improves user privacy, as less data is available in the public blockchain about the user (see Chapter 7). However, it also limits the auditability of the transactions. The optimization of the transactions before storing in the blockchain, also known as summarization, is proposed in [13, 14]. Recall from Chapter 3 that Catena [14] proposed a logging system where the transactions generated by the users are summarized in one transaction by the log server and only the summarized transaction is stored in the blockchain. Similarly, in [13], we proposed a solution to summarize transactions that is outlined in greater detail in Section 5.3.

The author in [15] proposed MimbleWimble, which enables the Bitcoin users to consolidate transactions that are stored in the blockchain into one transaction as shown in Figure 5.3. The MimbleWimble transactions are derived from confidential transactions, which are explained next. Confidential transactions aim to increase the privacy of the Bitcoin users by hiding the inputs and outputs in a transaction. To do so, homomorphic encryption is employed to encrypt the inputs/outputs using a blinding factor that essentially is a secret string known to the transaction generator and recipient. This potentially hides the input/output values from the Bitcoin users and thus enhances the privacy of the involved parties in the transaction. To verify a confidential transaction, the total number of inputs to the transactions must equal with the total number of outputs (e.g., transaction *x* with three inputs must have three outputs to be considered as a valid transaction).

MimbleWimble makes use of confidential transactions and enables the users to consolidate the transactions in a block in a single combined transaction (say, T_c). The validators in the network generate T_c where the input and output values of T_c are the unspent inputs and outputs of the transactions in the block. As an illustrative example, assume that Alice pays 1 BTC to Bob. Bob then pays 1 BTC to Chris. The input of the combined transaction is 1 BTC from Alice and the output is 1 BTC to Chris. Thus, in MimbleWimble, each block contains the following:

- *Block header:* This is as conventional blockchains.

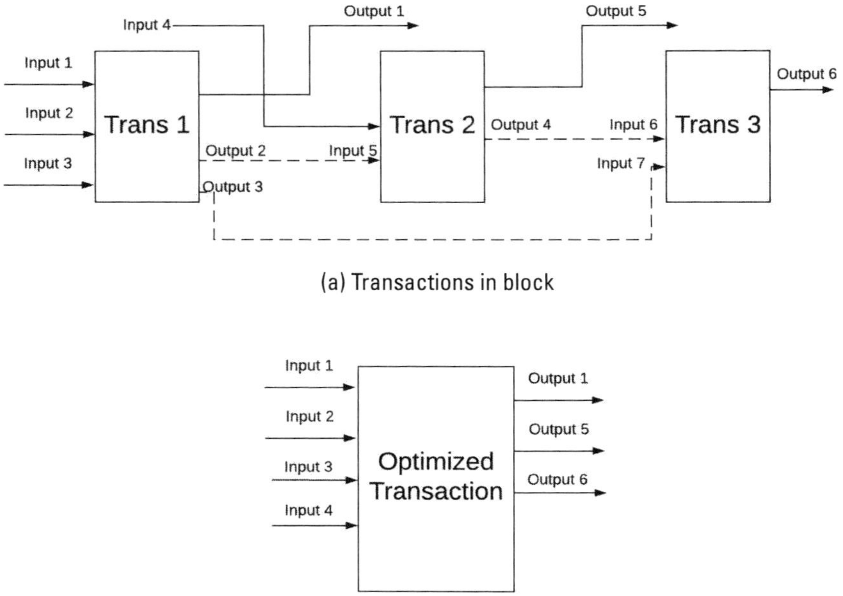

(a) Transactions in block

(b) Optimized transactions

Figure 5.3 An illustrative example of transaction optimization.

- *Inputs:* These are the inputs that are not spent by other transactions in the block (see Figure 5.3).
- *Outputs:* These are the outputs that are not spent by other transactions in the block (see Figure 5.3).
- *Kernel offset:* This is the blinding factor included in the blocks and prevents malicious nodes from reconstructing the input/output values.

If the size of the blockchain is 80 GB, it can be reduced to 30 GB by using MimbleWimble. MimbleWimble is employed by two cryptocurrencies: Grin and Beam. Each node in the network runs the combination function, as outlined above, on its blockchain version to reduce the blockchain size. However, a node may choose not to run this algorithm, which potentially increases the size of its database.

Palai et al. [16] proposed a blockchain that summarizes the transactions in blocks and thus reduces the blockchain size. To protect against a double-spending attack and to empower the users to verify newly generated transactions, the latest m blocks in the ledger must not be summarized where m is defined based on application. To prevent frequent summarization of the blocks, the remaining blocks in the chain are summarized when the cumulative number of the latest unsummarized blocks reaches a predefined value known as *l*. The validators collect *l* blocks and start the summarization process. Similar to MimbleWimble [15], the inputs/outputs of the summarized block are the inputs/outputs of the constitute transactions that are not spent by other transactions in the same block (see Figure 5.3). However, in MimbleWimble, it is impossible to audit the actual transaction corresponding to each input/output in the summarized block. To address this challenge, the authors in [16] proposed to tag the outputs/inputs in the summarized block with <Hash, transaction> where hash is the hash of the transaction. The hash of the summarized blocks (i.e., *l* blocks) is also stored in the summary block.

This section has discussed state-of-the-art approaches for reducing blockchain size and storage overhead, including mechanisms to store data off-the-chain and provide users with more control over their data, removing off-chain data, modifying the blockchain through Chameleon hash functions, and optimizing transactions through summarization. Each of the above optimizations has been proposed in isolation. How to have a versatile blockchain framework that allows a broad range of optimizations without affecting blockchain consistency is not addressed by these approaches. The next section details one proposal for a comprehensive framework for blockchain memory optimization.

5.3 A Memory-Optimized and Flexible Blockchain (MOF-BC)

In the previous section, we outlined some of the existing solutions that attempt to mitigate the impact of challenges associated with the immutability feature of blockchain. However, the existing solutions only partially address the outlined challenges in Section 5.1, which highlights the need for a comprehensive solution designed based on CPS requirements. In this section, we discuss a comprehensive solution to address the outlined challenges known as MOF-BC [13]. MOF-BC is a blockchain extension that enables users to remove transactions from the blockchain while maintaining the consistency of the chain. MOF-BC creates a separate layer on top of the blockchain that makes it independent of the underlying blockchain protocols (e.g., consensus algorithm) and thus can be implemented on top of any existing blockchains.

MOF-BC enables the participants to remove transactions in two modes: user-initiated memory optimization (UIMO) and network-initiated memory optimization (NIMO). In the former, the blockchain users initiate the transaction removal by generating a request to remove particular transactions, while, in the latter, the users delegate the network to remove their transactions under specific circumstances. MOF-BC introduces a number of different agents that assist in the removal of transactions in the NIMO mode and are discussed in the rest of this section. Similar to validators, the agents are selected among devices with high-resource availability that choose to function as agents. To protect against a single point of failure and enhance the security of the network, multiple replicas of each agent exist in the network that synchronize information among each other.

In conventional blockchain instantiations, the user pays a transaction fee to the validator that stores its transaction by following the consensus algorithm. The transaction fee incentivizes the nodes to dedicate resources and function as validators. Recall from Section 5.1 that the large volume of transactions in CPS may lead to an ever-increasing blockchain database size. This increases the cost in storing the blockchain database by the participating nodes as they must dedicate storage to store the full history of the blockchain. However, in the existing blockchain frameworks, such cost is neglected. To address this challenge, MOF-BC defines a storage cost that is periodically paid to the nodes in the network that store the blockchain. The storage cost is levied based on the size of the transactions and the duration for which the transaction is going to be stored in the blockchain (see discussions in Section 5.3.2.2). Assume that $|x|$ denotes the size of string x. To calculate the storage cost, MOF-BC considers the ratio between the actual transaction size and the size of the standard transactions in the blockchain denoted as $|page|$. As outlined in Chapter 3, normal blockchain transactions contain the PK$^+$ of the transaction generator and its corresponding signature, and the ID of the current and previous transactions shown as *T_ID* and *P_T_ID*, respectively. Thus, the size of a standard transaction can be measured as $|page| = |PK^+| + |Signature| + 2^*|T\_ID|$. Recall that the size of the output of a hash function is unique and independent of the input size; thus, we multiple the size of *T_ID* by 2 to consider both $|T\_ID|$ and $|P\_T\_ID|$. Depending on the application, $|page|$ may be different which can be defined by the blockchain designers.

The users must pay the storage fee upfront when generating a transaction. The fee is collected by a storage manager agent (StMA) that is responsible for collecting the storage fees and paying the same to the nodes that have stored the blockchain in particular time intervals known as payment interval. Each node that aims to dedicate storage space to the blockchain informs the StMA of its intention and the amount of storage that it is going to dedicate to the blockchain. The StMA adds the node to the list of the nodes that currently stored the blockchain. At the end of a payment interval, the StMA calculates the share of each node (say, node x) from the total amount of collected fees using the following equation:

$$Share_X = Storage_x * \frac{Fee}{Store_{All}} * \frac{Time_x}{PaymentPeriod} \tag{5.1}$$

where *Fee* represents the total storage fees collected by the StMA and $Time_x$ denotes the duration in which node x has stored the blockchain database.

Malicious nodes may falsely claim that they have stored the blockchain to receive the storage fee shares. To protect against this attack, MOF-BC introduces patrol agents (PAs) that are software agents migrating between the nodes. The PAs randomly migrate between the nodes in a way that during the payment interval they visit each node at least once. The random selection of the PAs increases the security against malicious nodes that may learn the pattern that the PAs visit and store the blockchain before the PA visit. Additionally, the possibility of a PA visiting a node more than once increases the security against malicious nodes that remove the blockchain after the PA visit until the next payment interval. If any inconsistency between the claim of the nodes and the actual storage space is detected, the PA informs the StMA. The latter then removes the corresponding node from the list of nodes that stored the blockchain.

5.3.1 Transaction Removal

In conventional blockchains, the hash of a block that constitutes n transactions is calculated as follows:

$$H(block) = H(header + Tran_1 + Tran_2 + \dots + Tran_n)$$

The *Tran* field contains all the fields in the transactions and thus makes it impossible to remove transactions from the blockchain without breaking the consistency. To address this challenge, MOF-BC calculates the hash of a block using the following:

$$H(block) = H(header + T_1\_ID + T_2\_ID + \dots + T_n\_ID)$$

Recall from Chapter 3 that *T_ID* is the hash of all the fields in the transaction that protect the integrity of the transaction content. To remove a transaction, the nodes remove the transaction content from the blockchain while maintaining the *T_ID*. The persistence of the *T_ID* enables recalculation of the block hash and thus protects the blockchain consistency and a level of auditability for the removed transactions in a way that the transaction generator can prove the existence of its transactions in the blockchain, but needs to maintain a copy of the transaction locally. To verify the existence of the transactions, the nodes need to hash the received transaction content and compare with the *T_ID* stored in the blockchain. A match can confirm the existence of the transaction in the blockchain.

5.3.2 Memory Optimization

In this section, we outline the details of different memory optimization methods employed in MOF-BC. To enable the users to remove their transactions, MOF-BC adds the following fields in the transactions:

$$GV \| MOM \| MOM\text{-}Setup \| Pay\text{-}by\text{-}reward$$

GV is employed to prove ownership of transactions, discussed in detail in Section 5.3.2.1. The *MOM* and *MOM-setup* fields identify the type of optimization and the optimization setting and are outlined in Section 5.3.2.2. Recall that MOF-BC introduces user-initiated memory optimization (UIMO) and network-initiated memory optimization (NIMO) modes that are discussed in the rest of this section.

5.3.2.1 User-Initiated Memory Optimization (UIMO)

The UIMO enables the users to remove a previously stored transaction in the blockchain. To remove a transaction (say, transaction *r*), the user (say, Alice) must first prove that she has previously generated *r*. This requires Alice to sign the remove transaction, that is, the transaction that requests the nodes to remove *r*, with PK^- corresponding to the PK^+ of *r*. Recall that the users may employ a fresh PK^+ for each new transaction to increase their anonymity. Additionally, the users may possess multiple devices that potentially lead to a significant number of transactions and thus key pairs. The storage of the keys incur significant processing and storage overhead and thus makes key management challenging. MOF-BC introduces the concept of Generator Verifier (GV) that is a signed hash of a secrete value that is unique in each transaction. The GV is calculated as: $GV = GV\text{-}PK^- \left(H(H(GVS) \| P\_T\_ID)\right)$, where GVS (generator verifier secret) is a secret only known to the user and can be any value (e.g., a string similar to a password or the biometric information of the user). The user may change the GVS value using a particular pattern to make the GVS in each transaction unique. Given that the hash output changes with small changes in the input, the pattern can be simply adding a static value to the GVS (e.g., a user may add 1 to the GVS as the pattern). $GV\text{-}PK^-/GV\text{-}PK^+$ is a unique key pair used by the users to sign the GV values in their transactions. The P_T_ID makes GV unique in each transaction that introduces the same level of anonymity for GV as using a fresh PK^+ in conventional blockchains.

Alice populates the P_T_ID, the hash of GVS, and $GV\text{-}PK^+$ in a transaction (say, X) and signs with $GV\text{-}PK^-$. Other nodes must first verify the GV as summarized in Figure 5.4. The first step is to decrypt the GV in *r* using the provided $GV\text{-}PK^+$. The resulting hash must match with the hash calculated by $H(H(GVS) \| P\_T\_ID)$ in the received transaction. To ensure that the generator of the transaction knows the GV-PK the signature in the transaction must

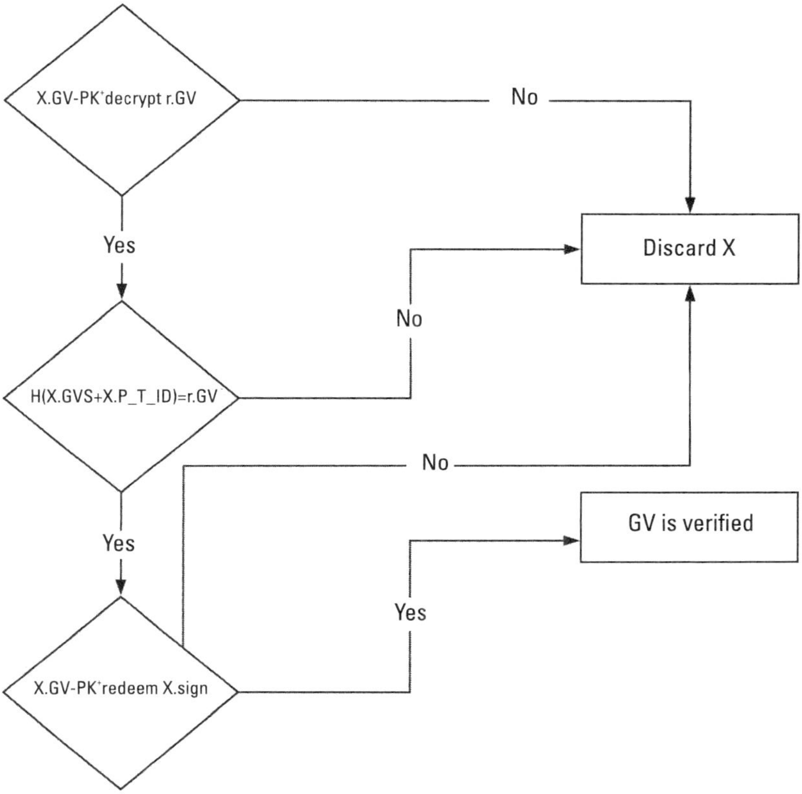

Figure 5.4 The process of verifying GV.

be verified with GV-PK$^+$. A match verifies the ownership of Alice. Thus, GV enables the users to manage all their transactions using a single key pair (i.e., GV-PK$^+$/GV-PK$^-$) and a secret value (i.e., GVS).

The MOF-BC defines three optimization modes in UIMO: transaction removal, summarization, and aging:

- *Removal:* To remove *r*, Alice generates a remove transaction that is structured as follows:

  $$T\_ID \| P\_T\_ID \| Input \| GV\text{-}Hash1 \| GV\text{-}Hash2 \| GV\text{-}PK^{+} \| Sign$$

  where *Input* is the ID of the transaction to be removed, and the next three fields relate to the GV verification as outlined above. The last field is the signed hash of the transaction using the GV/PK$^-$. The remove transaction is broadcast and stored in the blockchain. The participating nodes verify the ownership of the remove transaction generator as outlined earlier in this section. Next, the nodes that store the blockchain remove the transaction as outlined in Section 5.3.1.

The MOF-BC incentivizes the users remove their transactions from the blockchain database by awarding rewards for removing their transactions based on the space that they saved in the blockchain. A dedicated reward manager agent (RMA) collects remove transactions from the newly generated blocks and calculates the reward for each user based on the following:

$$Reward = Y.pages - X.pages$$

where Y is the remove transaction and X is the transaction that is removed from the blockchain. The RMA sends the total amount of reward accrued by a user along with the GV in the transaction to a bank that centrally manages the reward payment. The user can spend the collected rewards in two ways: (1) exchange with Bitcoin, in which the user can ask the bank to convert the rewards to Bitcoin with specified rate, and (2) pay the transaction fee, in which the collected rewards by the user can be spent as a storage fee in newly generated transaction. To inform the validators that the storage fee is paid by the accrued rewards, the user sets the Pay-by-Reward field in the newly generated transaction as 1. The user populates a packet with *T_ID* of the transaction that is going to be paid by reward along with the information needed to verify GV to bank. On receipt, the bank verifies GV and pays the storage fee of the identified transaction to the StMA. The bank then notifies the validators of *T_ID* of the transaction that is paid by reward.

- *Summarization:* As outlined in Section 5.1, a user may progressively install devices and thus may need to summarize the transactions in a single consolidated transaction. The summarization of transactions reduces the memory footprint of transactions in the blockchain and increases the user privacy. To summarize a group of transactions, Alice generates a summary transaction that is structured as follows:

$$Time\ Stamp \| PK^{+} \| Sign \| Merkle\ tree\ root \| Inputs \| Outputs \|$$
$$Summary\text{-}time \| Trans\text{-}order \| H_{x.1} \| H_{x.2} \| GV\text{-}PK^{+}$$

where *Time Stamp* is the time when the transaction was generated and PK^{+} and *Sign* are populated by Alice (i.e., summary transaction generator). *Merkle tree root* is the root of the Merkle tree constructed using the IDs of the transactions to be summarized (see Chapter 3 for a detailed discussion on a Merkle tree). The summarized transactions may contain inputs and outputs. To reduce the size of the summary transaction, only the inputs/outputs that are not used by other summarized transactions are included in the summary transaction. Conceptually, this is similar to MimbleWimble protocol [15] as outlined in Section 5.2.4. *Summary-time* contains the time when each summarized transaction was generated to provide a level of auditability. *Summary-time* stores the time stamp of the constitute transactions in order; however, the Merkle tree does not enforce any particular order. To address this challenge, the smallest number of distinct bytes of *T_ID* of the summarized transactions is stored in order in the *Trans-order* field. As an example, the *Trans-order* field for the transactions shown in Figure 5.5 is (vai2, bwpQ, vaiw, evas). To identify the correct order of a transaction, one should find the bytes in the *Trans-order* field that match with the first bytes of the transaction. The order of the matched bytes in *Trans-order* represents the order of the transaction. The last three fields are used to construct verify the GV for each of the summarized transactions as outlined earlier in this section.

Alice populates the summary transaction and broadcasts it to the entire blockchain network. The validators first verify if Alice is the owner of all transactions marked to be summarized using GV as outlined earlier in this section. If so, the validators store the transaction in the blockchain. The nodes that stored the blockchain then locate the transactions listed in the summary transaction and remove all as outlined in Section 5.3.1. The RMA calculates the reward for the summary transactions as follows:

$$\text{Reward} = \sum_{i=1}^{k} \text{ti.pages} - \text{Sum.pages} \tag{5.2}$$

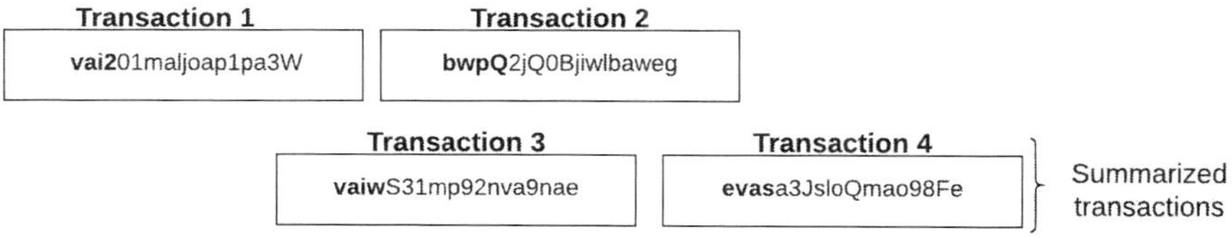

Figure 5.5 An example of the population of the Trans-order field.

where k is the total number of summarized transactions and *Sum* is the summary transaction.

- *Aging:* Recall that the transactions stored in the blockchain may either contain data or a link to data stored in a storage. In the latter, the modification of the data is not possible as the hash of the modified data will not match with the hash stored in the transaction. However, CPS devices generate a huge volume of data (e.g., videos or photographs), which potentially demands a significant storage space. To save space, the user may decide to compress the data in the storage, which potentially will change the hash and thus break the consistency.

  The MOF-BC empowers the users to age the data linked with transactions in the blockchain. To do so, Alice populates an *Age transaction* that is structured as follows:

$$T\_ID \| P\_T\_ID \| Original\_ID \| Content \| PK^{+} \| Sign \| H_{x.1} \| H_{x.2} \| GV\text{-}PK^{+}$$

  where *Original_ID* is the ID of the previous transaction in the blockchain that contains the link to the original data. The aged data is stored in the *Content* field. The rest of the fields are as discussed in summary transaction. Once the age transaction is stored in the blockchain, the nodes that stored the blockchain database remove the original transaction from their copy of the blockchain.

The outlined optimization modes are initiated by the user and allow them to manage their previously stored transactions in the blockchain. The UIMO requires the users to manage the memory optimization tasks for each of their transactions that potentially incurs a significant overhead on the users. To reduce the load, the MOF-BC introduces NIMO where the users can offload the memory optimization tasks to the network.

5.3.2.2 Network-Initiated Memory Optimization (NIMO)

In NIMO, the optimization task is initially authorized by the user and is conducted by the network participants in a distributed manner. To ensure that a user exerts control over their data, the user must identify the optimization type and settings during transaction generation. Recall from Section 5.3.2 that the MOF-BC introduces *MOM* and *MOM-Setup* fields in the transactions. The former identifies which MOM is to be used, while the latter identifies specific situations under which the transactions can be removed (e.g., an expiry time). These are explained in greater detail in the rest of this chapter. The MOF-BC introduces three memory optimization modes (MOM):

- *Temporary:* The transactions are stored in the blockchain for a particular period of time and are removed afterwards.
- *Permanent:* Similar to conventional blockchains, the transaction is permanently stored in the chain.
- *Summarizable:* The transaction can be summarized with other transactions in the same ledger.

Temporary transactions enable the users to store transactions for a particular period of time, known as the time to live (TTL). The TTL is stored in the *MOM-Setup* field in the transaction. Once the TTL expires, the nodes remove the transaction from their blockchain version. Note that, as the *MOM* and *MOM-Setup* fields are part of the transaction that is signed by the user, the GV field is set as null. Recall from Section 5.3 that MOF-BC introduces a storage fee that is paid to the nodes that store the blockchain database and varies based on the size and the duration for which the transaction is stored in the blockchains. The storage fee for temporary transactions is calculated as:

$$StorageFee = |Pages| * TTL * Rate \tag{5.3}$$

where *Pages* is the total number of pages required to store the transaction (see Section 5.3 for a detailed discussion), and *Rate* is the cost for storing a page for a particular period of time and is identified by the blockchain managers.

Similar to the conventional blockchains, the permanent MOM stores a transaction permanently. For permanent transactions, $StorageFee = |Pages| * Rate$. The rate is higher than the rate applied to the temporary transactions given that the transactions are permanently stored.

Recall from UIMO that users may consolidate a group of their transactions into a single summary transaction. In NIMO, the user off-loads the summarization process to the network, which is handled by a summary manager agent (SMA). An SMA searches all newly stored blocks to find transactions with summarizable MOM. If Alice wants her transactions to be summarized by the network, she has to chain all those transactions in a single ledger and set MOM for all as summarizable. The SMA collects all summarizable transactions generated during a particular time periods, known as summarization period, and at the end of the period generates a single summary transaction. The process of summarizing the transactions and the structure of the summary transaction are as outlined in UIMO. To protect against a malicious SMA that may attempt to generate a fake summary transaction, the validators that receive the summary transaction verify it before storing in the blockchain. The nodes first verify the signature of the SMA to ensure that the node that generated the transaction is a genuine SMA. Note that the PK^+ of the SMA is known to all participants in the network. Next, the nodes verify the Merkle tree root stored in the summary transaction by following the same steps as the SMA takes to generate the tree. If the root is verified, the transaction is considered valid and stored in the blockchain. Once the transactions are removed from the blockchain, the RMA calculates the reward as outlined for UIMO.

In this section, we outlined the process of removing transactions from the blockchain and different optimization models employed in the MOF-BC. To further reduce the associated overheads with the removing of the transactions, in the next section, we discuss the batch removal of transactions.

5.3.4 Batch Removal of Transactions

To remove a transaction, the nodes must first locate the transaction in the blockchain database, which incurs processing overhead, particularly in large-scale networks like CPS where a node may need to remove millions of transactions. This potentially keeps the nodes busy with the removing process and impacts their functionality as validators (or any other rule). To reduce the processing overheads, the nodes remove transactions at the end of particular time periods knowns as a cleaning period (CP). During the CP, the nodes function as normal and just store blocks in the blockchain. At the end of the CP, the nodes search their blockchain copy to locate the transactions that can be removed. The summarization period can be aligned with the CP to further reduce the processing overhead in the SMA.

Locating and removing all removable transactions in a block incurs overhead on the participating nodes. To reduce this overhead, the MOF-BC introduces a service agent (SA) that searches all newly generated blocks for removable/summarizable transactions. At the end of each CP, the SA removes transactions from its own version of the blockchain. The SA blockchain version is publicly available to all nodes; thus, they can download the latest version from the SA. To protect against malicious SAs, the nodes randomly verify the blocks received from the SA. Malicious behavior of the SA is reported to the network and the SA is replaced.

A user may request to remove all transactions corresponding to a device, which are chained in the same ledger. As an example, the user may wish to remove all transactions associated with a broken smart thermostat. To do so, the user must create a remove transaction for each of the transactions in the ledger, which incurs significant overhead on both the user and the network. To address this challenge, the MOF-BC enables the users to remove a ledger from the blockchain using a single transaction. The user generates a *ledger remove* transaction that contains the PK^+ and signature corresponding to the genesis transaction in the ledger. Recall from Chapter 3 that the genesis transaction is the first transaction in each ledger. The validators verify the signature and PK+ and store the transaction in the blockchain. During the next CP, the nodes that stored the blockchain remove all the transactions that are chained to the ledger.

The CP directly impacts the blockchain size. We measure this impact by implementing the proposed framework using C++ on a MacBook laptop (8 GB RAM, Intel core M-5y51 CPU, 1.10 GHz*4). We populated a blockchain with transactions generated by 900 nodes each generating one transaction per week. Each node generates different types of transactions: permanent transactions, temporary transactions where the TTL value for these transactions is 26, 52, 104, and 208 sequentially, and summarizable transactions. The implementation result is presented in Figure 5.6. As evident, the total amount of saved space increases with larger CPs. Recall that at the end of each CP the SMA consolidates summarizable transactions in one summary transaction, which impacts the size of the blockchain. Shorter CPs results in a larger number of summary transactions, which increases the size of the blockchain.

The larger CP increases the amount of saved storage; however, using large CPs incurs overhead on the validators to store expired transactions that is referred to as delayed optimization memory. Figure 5.7 outlines the implementation results to study the impact of CP on delay memory optimization. As evident, the larger CP incurs larger delay memory overhead, which wastes the resources of the validators. Thus, it is critical to consider the trade-off between the CP and the delayed memory space saved.

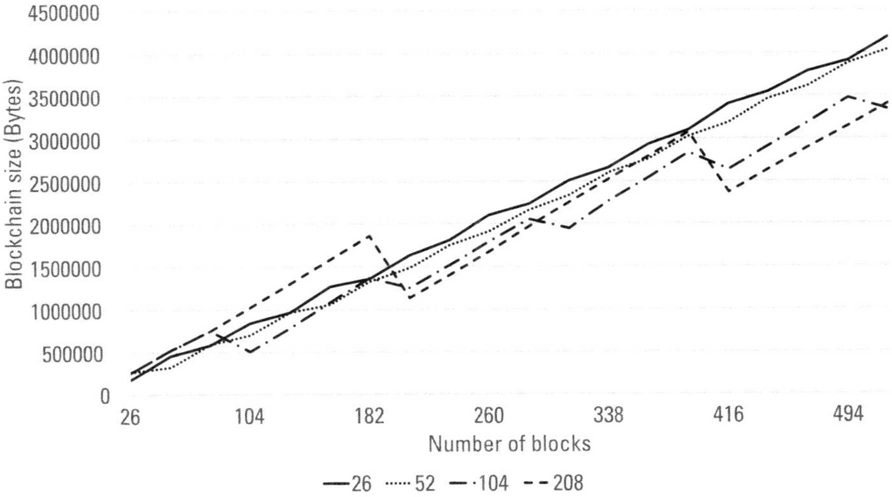

Figure 5.6 Evaluating the impact of CP on the blockchain size.

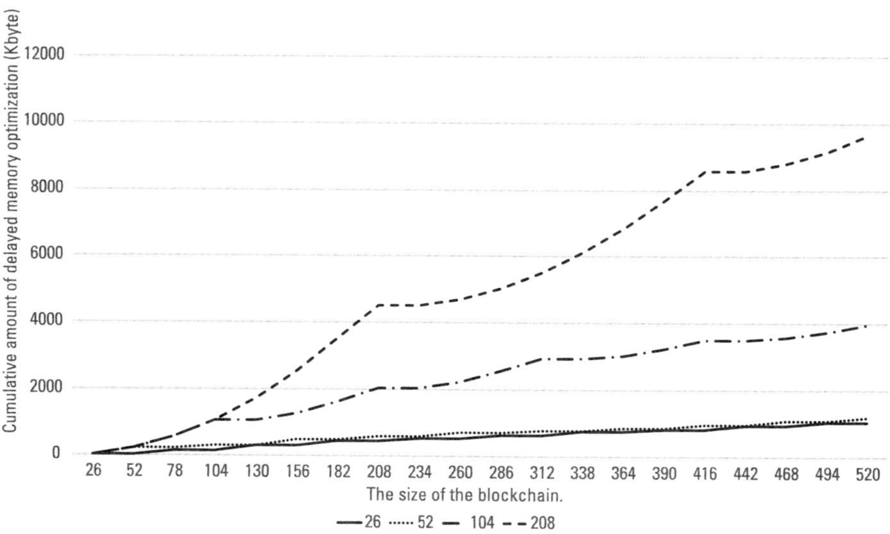

Figure 5.7 Evaluation of the impact of CP on the delayed memory optimization.

5.4 Comparative Evaluation

In this chapter, we studied the state-of-the-art solutions proposed to mitigate the impact of various challenges associated with blockchain immutability. We summarize the key methods and features in Table 5.1. We studied the compatibility of the discussed methods with five CPS requirements: (1) storage size, the degree in which the studied method reduces the size of the blockchain storage compared to the conventional chains; (2) user data exposure, the volume of data, including transactions, about a user available in the public chain; (3) flexible transaction, the flexibility in storing transactions for limited time or the ability to summarize; (4) GDPR, the compatibility with GDPR requirements; and (5) data persistence, the ability to remove information from the blockchain database. We categorize their compatibility in three groups: high, poor, and medium, where high has the highest compatibility with CPS and thus the most suitable option.

It is clear from Table 5.1 that each method addresses only some of the CPS requirements and that the strengths of individual methods can be complimentary. The MOF-BC, as a comprehensive and versatile memory-optimized blockchain, combines the benefits of individual methods to meet all CPS requirements.

Table 5.1
A Summary of the State-of-the-Art Solutions to Optimize Memory Optimization

		Compatibility with CPS Requirement				
Optimization Method	**Features**	**Storage Size**	**User Data Exposure**	**Flexible Transaction**	**GDPR**	**Data Persistence**
Off-chain storage	Reduces the size of blockchain by storing data off-the-chain	Medium	Medium	Poor	Poor	Poor
Removing off-chain data	Enables the users to remove data that is stored off-the-chain while maintaining the blockchain consistency	Medium	Medium	Poor	Medium	Poor
Data modification	Enables the selected nodes to modify the content of a block	High	High	Poor	Medium	High
Optimizing transactions	Reduces the size of the blockchain by removing the spent input/outputs	Medium	Medium	Poor	Poor	Poor
MOF-BC	Reduces the size of the blockchain by offering multiple memory optimization modes including summary, temporary and aging; enables the users to remove transactions from the blockchain while maintaining consistency	High	High	High	High	High

5.5 Summary

Blockchain immutability makes it virtually impossible to remove or modify transactions in the chain. However, in large-scale networks like CPS, this raises challenges including storage size, privacy, cost, and compliance with regulations. We saw in this chapter that several proposals have emerged for challenging the common perception of blockchain immutability without affecting the blockchain consistency and auditability. Our review of state-of-the-art methods identified four categories of memory optimization approaches in blockchain, namely, off-chain data storage, removing off-chain data, data modification, and optimizing. While each of these methods carries its own advantages for large-scale networks like CPS, none of them can individually cover all of the CPS requirements for memory optimization that we identified at the beginning of this chapter. In the latter part of the chapter, we discussed a comprehensive solution to address the storage challenges in the blockchain also known as the MOF-BC in Section 5.3. The MOF-BC supports all types of memory optimizations for blockchain, is blockchain protocol-agnostic, and incorporates incentives for reducing blockchain storage space. Open challenges in this space include the deeper exploration of the trade-offs involved in individual and composite memory optimization methods, such as the cleaning period duration in the MOF-BC, in terms of bandwidth, computation, and storage for different context, to guide application-specific memory optimization in different blockchain scenarios.

In this chapter, we studied the challenges associated with blockchain immutability. As outlined in Chapters 2 and 3, establishing trust among the participants in a decentralized system is challenging. In the next chapter, we study trust management in blockchain.

References

[1] "Blockchain Charts," https://www.blockchain.com/charts/blocks-size.

[2] "Data Protection Reform Package," https://ec.europa.eu/commission/presscorner/detail/en/MEMO_17_1441.

[3] Zyskind, G., and O. Nathan, "Decentralizing Privacy: Using Blockchain to Protect Personal Data," *2015 IEEE Security and Privacy Workshops*, May 2015, pp. 180–184.

[4] Shafagh, H., et al., "Towards Blockchain-Based Auditable Storage and Sharing of IoT Data," *Proceedings of the 2017 on Cloud Computing Security Workshop*, November 2017, pp. 45–50.

[5] Theodouli, A., et al., "On the Design of a Blockchain-Based System to Facilitate Healthcare Data Sharing," *2018 17th IEEE International Conference On Trust, Security and Privacy in Computing and Communications/12th IEEE International Conference on Big Data Science and Engineering (TrustCom/BigDataSE)*, August 2018, pp. 1374–1379.

[6] "Bitcoin Version 0.11.0," https://bitcoin.org/en/release/v0.11.0.

[7] Bayle, A., et al., "When Blockchain Meets the Right to Be Forgotten: Technology Versus Law in the Healthcare Industry," *2018 IEEE/WIC/ACM International Conference on Web Intelligence (WI)*, December 2018, pp. 788–792.

[8] Makhdoom, I., et al., "PrivySharing: A Blockchain-Based Framework for Privacy-Preserving and Secure Data Sharing in Smart Cities," *Computers & Security*, Vol. 88, 2020, p. 101653.

[9] Rieger, A., et al., "Building a Blockchain Application That Complies with the EU General Data Protection Regulation," *MIS Quarterly Executive*, Vol. 18, No. 4, Article 7, 2019.

[10] Al-Zaben, N., et al., "General Data Protection Regulation Complied Blockchain Architecture for Personally Identifiable Information Management," *2018 IEEE International Conference on Computing, Electronics & Communications Engineering (iCCECE)*, August 2018, pp. 77–82.

[11] Ateniese, G., et al., "Redactable Blockchain–or–Rewriting History on Bitcoin and Friends," *2017 IEEE European Symposium on Security and Privacy (EuroS&P)*, April 2017, pp. 111–126.

[12] Krawczyk, H. M., and T. D. Rabin, ""Chameleon Hashing and Signatures," U.S. Patent No. 6,108,783. Washington, DC: U.S. Patent and Trademark Office, 2000.

[13] Dorri, A., S. S. Kanhere, and R. Jurdak, "MOF-BC: A Memory Optimized and Flexible Blockchain for Large Scale Networks," *Future Generation Computer Systems*, Vol. 92, 2019, pp. 357–373.

[14] Tomescu, A., and S. Devadas, "Catena: Efficient Non-Equivocation Via Bitcoin," *2017 IEEE Symposium on Security and Privacy (SP)*, 2017.

[15] "Mimblewimble," https://download.wpsoftware.net/bitcoin/wizardry/mimblewimble.txt.

[16] Palai, A., M. Vora, and A. Shah, "Empowering Light Nodes in Blockchains with Block Summarization," *2018 9th IFIP International Conference on New Technologies, Mobility and Security (NTMS)*, February 2018, pp. 1–5.

6

Managing Data Trust in Blockchain

Volkan Dedeoglu
Data61, CSIRO, Brisbane, Australia

6.1 Introduction

With the exponential growth of CPS applications and their impact on businesses and society through devising added values for existing services or creating new ones, the notion of trust in CPS applications has emerged as a computational concept. CPS applications heavily rely on data collected by the sensors to capture observations of the physical domain and record them digitally. The process usually involves converting an analog signal captured from the physical domain into a digital signal, which is then recorded and processed according to the application logic. In other words, sensor devices provide the observations of the true state of the physical domain as an input to the application logic. However, these sensor devices usually have limited resources and capabilities and they are deployed in physical environments, which may degrade the quality of observations or expose the devices to a multitude of attacks. Thus, the observations may be subject to noise, bias, sensor drift, or malicious alterations.

The trustworthiness of a CPS application can be assessed in three distinct layers [1]:

- *Data layer* relates to sensor and other observational data.
- *Interaction layer* relates to communications among devices in the CPS.
- *Application layer* relates to data processing and the interactions between service providers and service users.

To establish trust in a CPS application, trust mechanisms should consider not only the performance of all three layers but also the reliable interaction among these layers and ensure the end-to-end integrity of the collected data and the associated interactions. Transparency and auditability of data collection processes and the associated interactions are key pillars for trust management in CPS. Recently, blockchain has been proposed as a technology that enables transparency and auditability in CPS applications.

Based on consensus mechanisms, such as POW [2], POS [3], or POET [4], and maintained by a distributed network, blockchain is a shared ledger that can record transactions in an immutable way. By establishing an agreement on the current state of the ledger among trustless network participants, the consensus mechanisms build a new dimension of trust, which is decentralized due to its coupling to the outcome of distributed consensus among the network participants.

As discussed in Chapter 4, for CPS applications, blockchain provides an immutable and auditable history of sensor observations and interactions among CPS devices and other network entities by recording them in blocks that are chained through cryptographic hash links to previous blocks in the blockchain. Cryptographic hash links and distributed consensus mechanisms ensure that it is virtually impossible to alter or delete the data stored on an immutable blockchain without detection. Furthermore, the authenticity of CPS transactions and blocks can be verified using public key cryptography. Blockchain mechanisms guarantee that the internode interactions recorded in the block's transactions are securely recorded and are tamper-proof, once the blocks are mined into the blockchain. However, a tamper-proof record of internode interactions is a necessary but insufficient element to deliver end-to-end trust in CPS.

Consider a CPS device whose observation is not accurate. This observational data can be input to a CPS application that stores the hash of the data in a blockchain for verification. Storing the hash of the data on the blockchain ensures the integrity of the stored data, which can be verified by comparing its hash against the blockchain-stored hash value. However, the blockchain-stored hash value cannot guarantee the authenticity of the observational data itself in the first place. Although the observational data is secured with blockchain, the data itself may be inaccurate and an application using this data may produce unwanted outcomes. The immutability of blockchain does not protect against this risk associated with data capture, and as a result, the data may not be useful to the CPS end users.

As discussed above, trust establishment in CPS has an intertwined nature requiring trust management for both the internode interactions and the data capture processes. Much of the existing literature fails to recognize this intertwined nature of the trust problem in CPS and tries to address the trust problem as two separate problems. Trust and reputation-based approaches have been introduced for establishing trust for internode interactions, and evidence-based approaches have been used for CPS data trust. Therefore, this chapter builds on the need for an integrated architecture that can establish end-to-end trust for CPS applications by considering both the data collection processes and blockchain node interactions.

In this chapter, we first study the data trust problem and the existing trust mechanisms in Sections 6.2 and 6.3, respectively. In Section 6.4, we introduce a multilayered architecture for improving the end-to-end trust that can be applied to a diverse range of blockchain-based CPS applications. The architecture evaluates the trustworthiness of observations at the data layer by using fusion of information from different sources, confidence of observations, and reputation level of data sources. At the blockchain layer, the architecture incorporates lightweight block generation, reputation-based adaptive block validation, and distributed consensus mechanisms. Section 6.5 presents the security analysis of a trust management framework, while Section 6.6 concludes the chapter.

6.2 Trust in CPS Applications

It is hard to define the concept of trust in the CPS context, which encompasses all the system components (e.g., data, devices, entities) and their complex interactions regarding reliability, truthfulness, security, confidence, experiences, and beliefs. Despite its perception as a trust machine, blockchain alone cannot guarantee end-to-end trust for CPS applications that rely on data collected by CPS devices. In this section, we introduce the trust notions associated with the data stored on blockchain and network participants.

6.2.1 Trusting the Data (Data-Centric)

Trust in the data refers to the belief about how likely the data represents the ground truth. The immutability of blockchain ensures that, once data is stored on a blockchain, it is very hard to modify or delete the data due to the structure of the hash links connecting the blocks and the consensus mechanisms that make block mining a difficult task. Thus, we trust that the data stored on blockchain cannot be altered. However, the mechanisms guaranteeing the immutability of blockchain do not guarantee the trustworthiness of data necessarily. In principle, we can trust the data stored on a blockchain as long as the data can be verified without needing any external input to the blockchain and the blockchain is trusted. We will start our trust discussion by examining the case for data generated on the blockchain and then move to the case for data generated outside the blockchain.

6.2.1.1 Data Native to the Blockchain

When data is generated on blockchain protocols, it can be verified by the transaction history recorded on the blockchain. The data generated on the blockchain can be considered internal, as the source of the data is the blockchain. The time-stamped transactions recorded on the blockchain are transparent and easily verifiable by other nodes in the network, which enables traceability. The trustworthiness of data generated on the blockchain is guaranteed by the blockchain mechanisms. Thus, we can trust the data native to the blockchain as long as the blockchain is built on trusted mechanisms.

As an example, consider the scenario depicted in Figure 6.1, in which Alice mines a new block on a blockchain and earns 25 coins as mining reward. Then Alice sends 10 coins to Bob, who initially had no coins. As a result of this transaction, Bob has 10 coins in his account and Alice has 15 coins remaining. All the data regarding to the coin amounts on Alice's account and Bob's account are recorded on the blockchain transactions, which can be verified by exploring the blockchain transaction history. Thus, when we check the accounts of Alice and Bob, we trust the data recorded on the blockchain.

6.2.1.2 Data Generated External to the Blockchain

Applications that require interactions with the world outside the blockchain may rely on data that is generated external to the blockchain. In this case, the data is pushed into a blockchain by a network node. Because the source of the data is not the blockchain, it cannot be verified by the blockchain transaction history. Although the blockchain ensures that the data cannot be tampered with once recorded on the blockchain, there is a trust problem related to the origin of the data, which is referred to as "garbage in, garbage out" problem [5].

As an example, consider a CPS application where data generated by sensors and collected by some gateway nodes is stored in a blockchain. Relying on the sensor inputs, smart contracts can be executed to process the data and perform predefined tasks. As the execution of the smart contracts depends on the data input, there is a need to guarantee that the data stored in the blockchain is authentic and reliable, in particular, when the sensor observations may

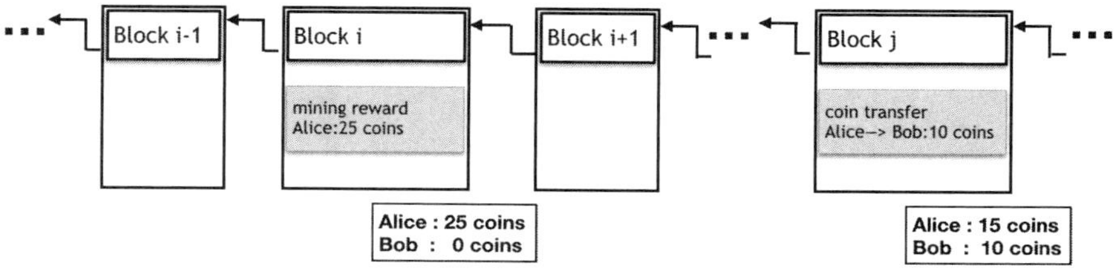

Figure 6.1 The data generated on the blockchain can be verified by the transaction's history.

be subject to noise, bias, sensor drift, or malicious alterations. Otherwise, the trustworthiness of the data cannot be verified on the blockchain, and false data may be recorded in the blockchain as depicted in Figure 6.2.

CPS applications heavily rely on data collected from the physical domain that is generated external to the blockchain. This distinction in the source of the data requires a data-centric approach to establish the trustworthiness of the data before it stored on the blockchain.

6.2.2 Trusting the Network Participants (Entity-Centric)

Trust in a network participant refers to the belief about how likely that participant will behave in an expected way. One of the main promises of the blockchain technology is establishing a new dimension of trust among a network of nodes without any intermediaries. The blockchain generates trust using some costly mechanisms, such as POW or POS, which is required when network nodes are anonymous, their identities are not known, and there is no existing trust relationship between them. However, in CPS applications, private or consortium network settings, network nodes may have known identities, which they use to register to the network. Furthermore, network participants may have continuing interactions, which may help them to establish reputations based on the history of interactions and experiences. Then, using an entity-centric approach, it is possible to establish some trust relationships among the

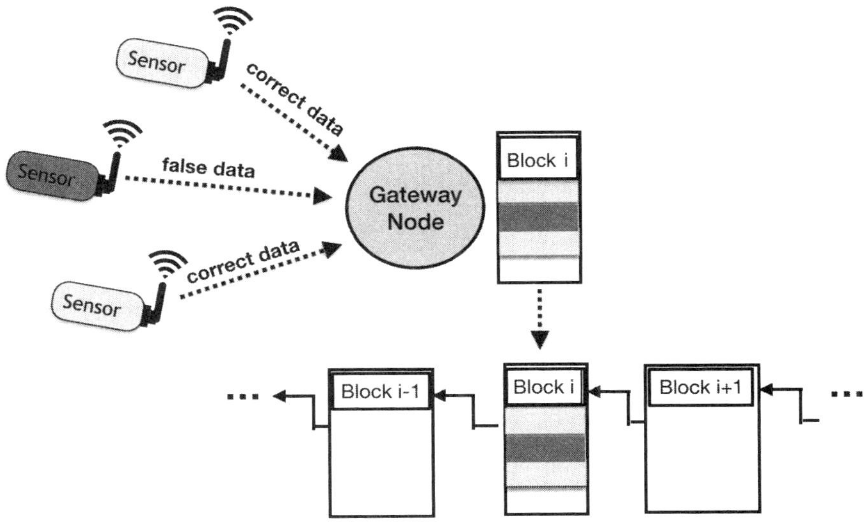

Figure 6.2 "Garbage in, garbage out" problem for data generated external to the blockchain. There is no guarantee that a sensor observation that is recorded on the blockchain immutably is also trustworthy.

network participants and to use this existing trust among them to improve the efficiency of blockchain mechanisms.

Having differentiated between data-centric and entity-centric trust, we discuss existing CPS trust management approaches next.

6.3 Existing Trust Management Approaches

Traditional trust management systems rely on the established trust relations among network entities to assess trust in CPS applications. This entity-centric approach uses the relationship among entities, the long-term behavior, experience, and knowledge of entities to attribute trust to data-reporting entities. Then the trust level of the data source is used to estimate the trust level of data. However, emerging CPS applications are heavily data-centric, and the entity-centric approach ignores the dynamic nature of the CPS data (e.g., timeliness, locality, confidence, validity) sourced by the entities. Thus, the trust level of entities may not represent the trust level in the data sourced by the entities and the entity-centric approach alone may not be sufficient for trust evaluation in data-driven CPS applications [6]. Therefore, there is also a need for data-centric trust assessment for CPS applications that takes into account the dynamic nature of the CPS data.

In [7], a data-centric trust establishment framework was introduced wherein the trust in data is derived based on multiple pieces of evidence. Based on multisensor data fusion techniques [8], the introduced scheme weighs these pieces of evidence by considering the dynamic aspects of the data (e.g., timeliness, locality) and the context of the CPS application to obtain an aggregate trust level for the data. Some of the techniques used for combining the evidence include majority voting, most trusted evidence, weighted voting, Bayesian inference, and the Dempster-Shafer theory [9]. In [10], distributed methods have been introduced to establish trust when multiple observer nodes within spatial or temporal proximity independently corroborate nearby observations.

The evidence for data can be captured by direct relations, indirect relations, or a combination of direct and indirect relations, as shown in Figure 6.3. While entities trust direct relations for evidence, the indirect relations can provide useful data as evidence when there is no direct relation or when the indirect evidence can complement the evidence provided by direct relations.

Due to the diverse range of emerging CPS applications and services, such as smart cities, smart homes, supply chains, healthcare, industry, energy, environmental monitoring, military, and transportation, it is hard to propose a generic evidence formulation. Based on the physical phenomena observed, and the heterogeneous nature of the available data sources, the existing evidence calculation schemes are application-specific.

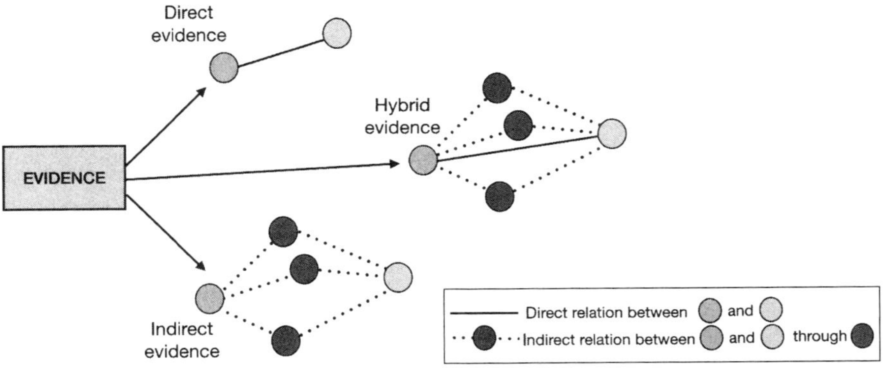

Figure 6.3 The evidence may be direct, indirect, or hybrid depending on the node interactions.

Consider a target detection application example as shown in Figure 6.4, where the evidence for the observation of node S_i can be the detection of the target by other sensor nodes in a predefined neighborhood N_i. As discussed before, there are different techniques for combining the evidence from the sensor nodes. One way to aggregate the evidence is the trust-weighted evidence calcu-

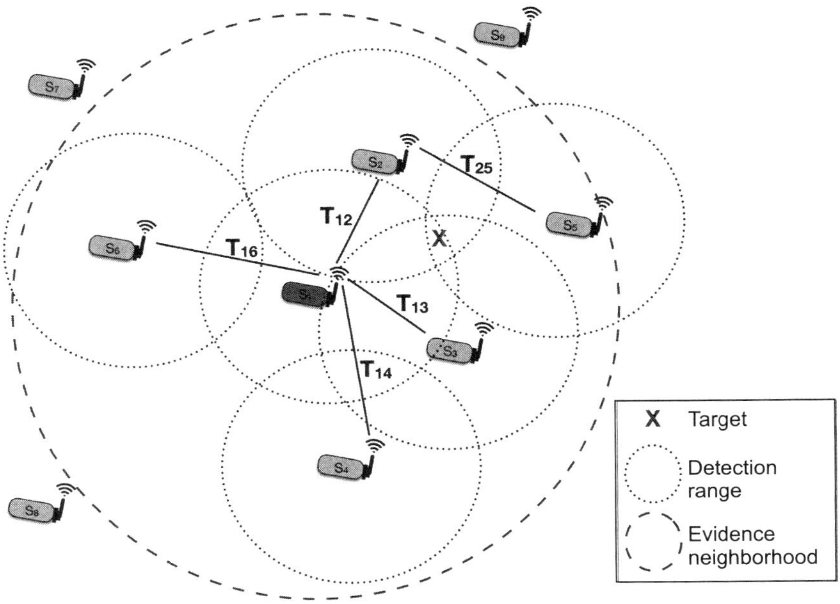

Figure 6.4 Evidence calculation for a target detection application based on the trust relationship among sensor nodes, where Tij is the trust-based evidence weighting factor between nodes Si and Sj.

lation. We can fuse the trust-weighted evidence for the observation of node S_i as follows:

$$\text{Evidence}(S_i) = \frac{1}{|\mathrm{N}_i|} \times \sum_{S_j \in \mathrm{N}_i} 1_i^j T_{ij} \tag{6.1}$$

where

$$1_i^j = \begin{cases} 1, & S_j \textit{ supports the observation of } S_i \\ -1, & \textit{otherwise} \end{cases}$$

and T_{ij} is the weighting factor calculated based on the trust relationship between nodes S_i and S_j. For the given example, sensor node S_1 has a direct trust relationship with nodes S_2, S_3, S_4, and S_6, and an indirect trust relationship with S_5. As shown in Figure 6.4, within the evidence neighborhood, nodes S_2, S_3, and S_5 detect the target and support the observation of S_1, whereas S_4 and S_6 do not detect the target and thus the evidence submitted by them does not support the observation of S_1. Nodes S_7, S_8 and S_9 are outside the evidence neighborhood, so their observations are not taken into account while calculating the evidence. Thus, the evidence for the observation of node S_1 can be calculated as:

$$\text{Evidence}(S_1) = \frac{1}{5} \times \left(T_{12} + T_{13} + T_{12} \times T_{25} - T_{14} - T_{16} \right) \tag{6.2}$$

Note that, due to the indirect trust relationship between nodes S_1 and S_5 through node S_2, the corresponding trust weight is calculated as $T_{12} \times T_{25}$. In Section 6.4.2.1, we provide another evidence calculation scheme based on the confidence of sensor observations.

The survey paper by Yan et al. [11] explored the properties that affect the CPS trust relationships and reviews the literature on CPS trust management mechanisms and their objectives. Recently, there has been an increasing interest in integrating the trust and reputation management mechanisms into blockchain-empowered CPS applications. As the blockchain provides a distributed platform for CPS applications, there is a need for decentralized trust and reputation management mechanisms for integration. Furthermore, due to the variety of blockchain-empowered CPS applications and services with different network topologies, rules of participation and governance, and node interactions, the existing proposals for establishing trust management are application-specific. In Table 6.1, we broadly categorize the existing trust management proposals based on the layer at which the trust mechanisms work as CPS data capture and blockchain node interactions.

In summary, the existing approaches for trust management in blockchain-based CPS applications cannot provide end-to-end trust as they either consider the data capture process for improving the trust in CPS data or the internode interactions for the nodes participating in the blockchain network. In the next section, we introduce our trust management architecture that takes into account both the data and the blockchain layers to improve the end-to-end trust for CPS applications.

Table 6.1
Trust Management for Blockchain-Based CPS Applications

	Research Work	Contributions
IoT Data Capture	Lu et al. [12]	Blockchain-Based Anonymous Reputation System (BARS), which uses direct historical interactions and indirect opinions about vehicles to establish a trusted communication environment for vehicular applications; their system determines the trust level of broadcasted messages based on the reputation score of the vehicles
	Kang et al. [13]	Reputation based high-quality data sharing scheme for vehicular networks using a consortium blockchain, smart contracts, and a subjective logic model, which relies on interaction frequency, event timeliness, and trajectory similarity for reputation management
	Kchaou et al. [14]	Distributed trust management scheme to calculate the credibility of exchanged messages based on the reputation value of observers in blockchain-based vehicular networks
Blockchain Node Interactions	Kang et al. [15]	Delegated Proof-of-Stake consensus scheme for secure data sharing in blockchain-enabled Internet of Vehicles; they used reputation-based voting to select the miners, where the reputation of the miner candidates is calculated utilizing a multiweight subjective logic scheme; they also presented a contract theory-based mechanism to incentivize the standby miners to participate in block verification
	Dorri et al. [16]	Lightweight Scalable Blockchain (LSB) for CPS with an IoT-friendly consensus algorithm that incorporates a distributed trust method to the block verification mechanism; the architecture has two tiers: overlay and smart home networks; based on direct and indirect evidence, the overlay network nodes build trust, which is used to reduce the number of transactions to be validated in a new block
	Bahri and Girdzijauskas [17]	Proof-of-Trust consensus mechanism, which capitalizes the existing trust between nodes to improve the energy consumption of POW consensus by adapting the difficulty level of POW for individual nodes depending on their trust measures

6.4 A Multilayered Trust Management Framework

In [18], we introduced an end-to-end trust architecture for blockchain-empowered CPS applications. Figure 6.5 depicts the three key layers of the blockchain-empowered CPS architecture and the two modules for the trust management framework:

- *Data layer:* The inputs of the CPS architecture constitute the data layer, which leverages IoT devices and other data sources to collect observational data, such as sensor measurements or manually entered data by regulators or social media streams. The data collected from different sources represent the observations of the physical world and can be stored on-chain or off-chain, while transactions recording its collection and communication are stored on-chain. If the observations are stored off-chain in a database due to scalability or cost-effectiveness, the hash of the observational data can be stored on-chain for data integrity and non-repudiation.
- *Blockchain layer:* The data (or hash of the data) received from the data layer is stored on a blockchain maintained by the blockchain layer. The blockchain layer acts as a distributed computing platform facilitating distributed data storage and data access in a secure, transparent, and immutable way. The blockchain layer also interacts with the application layer.
- *Application layer:* The application layer is responsible for processing the data collected by the data layer, while also providing an interface for the end users for communications and services. The application layer also interacts with the blockchain layer and controls the blockchain mecha-

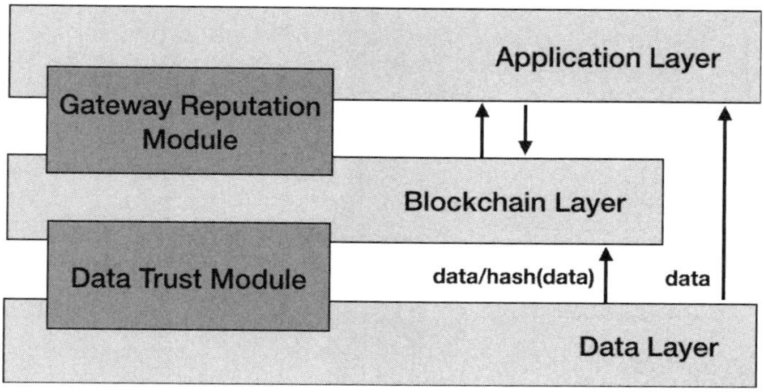

Figure 6.5 Multilayered trust architecture.

nisms based on the application-specific requirements and the specifications provided by the end users.

Our architecture introduces two key modules for end-to-end trust management that utilizes both data-centric and entity-centric approaches for trust establishment:

- *Data trust module:* As described in Section 6.2, CPS applications rely on the data collected from the physical domain. The data trust module evaluates the trustworthiness of this observational data using the confidence level of the observation reported by the data source, the evidence received from other data sources, and the long-time accumulated reputation of the data source (see Figure 6.6). Using the available observational data, confidences, and the current reputations of the data sources, the data trust module assigns a trust value for observations and records these trust values together with the associated observational data in transactions.
- *Gateway reputation module:* CPS applications are prone to security attacks or failures of network participants. Thus, to ensure end-to-end trust in CPS applications, we also need to consider the trustworthiness of the blockchain network participants. Figure 6.7 depicts the gateway reputation module that we propose to track the long-term reliability of the network participants. The gateway reputation module continuously evaluates and updates the reputation for blockchain participants based on the information that it receives from the blockchain network and external reputation sources, such as the reputation or transactions activity from external systems. The updated reputation scores are shared with the blockchain and application layers. As discussed in Section 6.2.2, the efficiency of blockchain mechanisms can be improved by considering the trustworthiness of the blockchain participants. The reputation scores calculated by the gateway reputation module can be used to

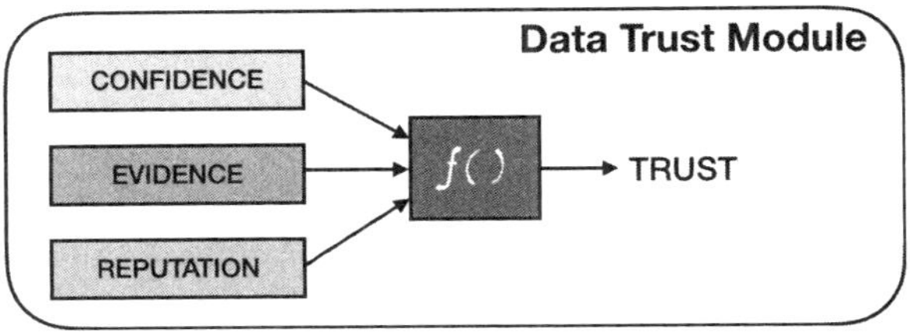

Figure 6.6 The data trust module evaluates the trustworthiness of an observation based on the confidence, evidence, and reputation components.

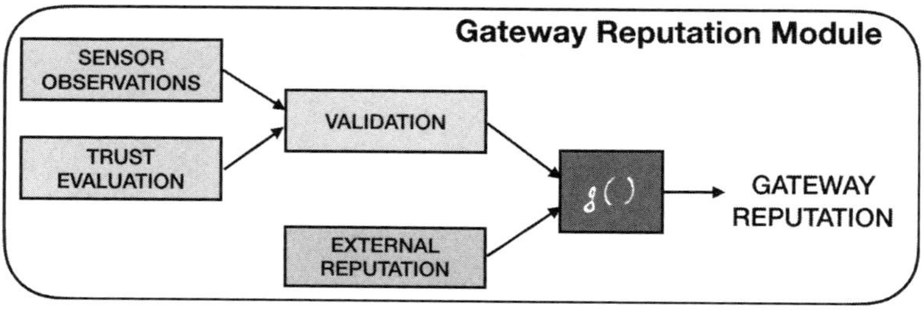

Figure 6.7 The reputation of a network participant depends on the validation by other network nodes and any external reputation transferred from other networks.

dynamically adapt transaction or block validation requirements of the participants. For example, if a participant has a high reputation score, the block generated by that participant receives less scrutiny, and the block validation mechanism can be adapted accordingly. This adaptation improves the efficiency and the performance of the blockchain. The dynamic reputation scores of the participants are also used by the incentive mechanisms at the application layer. For example, while the participants with high reputation scores can be offered economic incentives through increased business interactions, the participants with low reputation scores can be penalized.

6.4.1 Two-Tiered Network Model for CPS Applications

Before proceeding to the details of our end-to-end trust architecture, this section introduces the underlying network model for a generic CPS, which is composed of a set of gateway nodes and sensor nodes connected to these gateway nodes as shown in Figure 6.8. Since most CPS applications rely on sensor nodes, which are resource-constrained IoT devices with limited capabilities, we consider the two-tiered network model depicted in Figure 6.8. This application-agnostic two-tiered network model is a generic architecture for IoT networks [19] and can be applied to a diverse range of CPS applications.

- *Upper tier:* In the upper tier, we have a set of N gateway nodes $G = \{G_1, G_2, ..., G_N\}$. These gateway nodes are responsible for:
  - *Data collection from sensor nodes:* Without loss of generality, we assume that each gateway node G_i is associated with a set of K sensor nodes $S_i = \{S_{i1}, S_{i2}, ... S_{iK}\}$ at the lower layer. Gateway nodes collect the observational data from the associated sensor nodes.

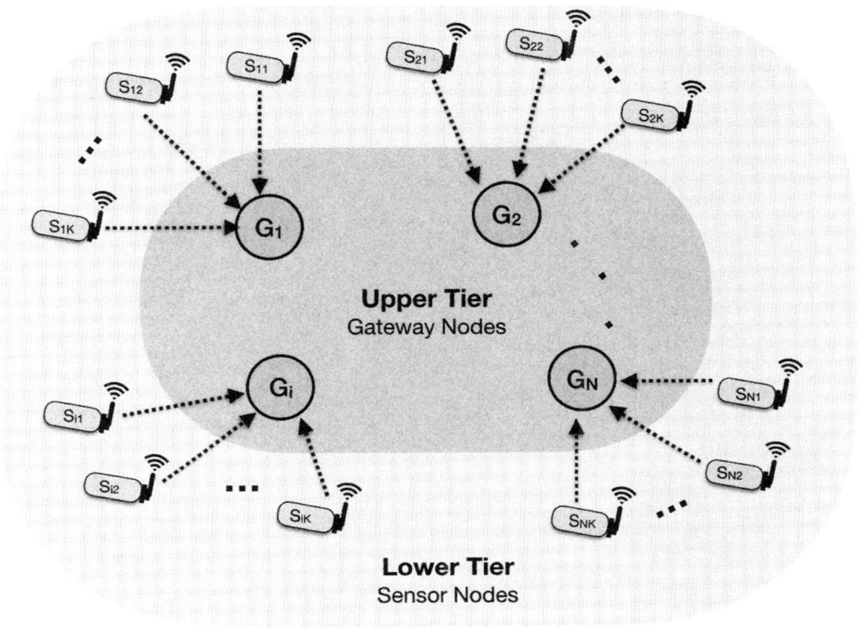

Figure 6.8 Two-tiered network structure. Data is collected by sensor nodes associated with gateway nodes, which maintain the blockchain overlay network.

- *Trust evaluation:* Once the observational data is collected from the associated sensor nodes, the gateway node evaluates the trustworthiness of the collected data using the trust management framework described in Section 6.4.2.1.
- *Blockchain network maintenance:* The gateway nodes maintain a blockchain network by participating in the block generation and block validation processes.

- *Lower tier:* In the lower tier, we have the resource-constrained sensor nodes with limited capabilities. These sensor nodes collect observational data from the environment and transmit the collected data to the associated gateway nodes through transactions. Since the sensor observations are prone to failures, noise, sensor drift, and security attacks, a large number of sensor nodes may be deployed densely to observe a physical phenomenon of interest in CPS applications. Due to this dense deployment and the spatiotemporal characteristics of the physical phenomenon to be observed, the sensor nodes associated with the same gateway and in close proximity to each other are assumed to have correlated observations (e.g., acoustic sensors in a room). For larger networks where a large number of nodes are associated with a gateway node, nodes can be clustered so that observations within the same cluster are highly correlated [20].

For authentication, every node in the network holds a unique public and private key pair (PK, SK).

- *Public key (PK):* During the network initialization phase, nodes are registered to the network with their public keys and digital profiles are created for the nodes on the blockchain to record their public keys, their network associations (i.e., the public keys of the associated nodes), and their reputation scores.
- *Private key (SK):* Nodes use their private keys to sign their transactions. These signatures can be verified by the gateway nodes, which have access to the public keys of the nodes recorded in their digital profiles.

6.4.2 End-to-End Trust Management Framework

As briefly discussed in the previous sections, our end-to-end trust management framework is built on a two-tiered network model and a layered application architecture. In this section, we describe in detail how we improve the end-to-end trust in CPS applications through the generic mechanisms for managing trust and reputation within this framework.

6.4.2.1 Trust Evaluation for Sensor Data

When a gateway node receives observational data from an associated sensor node, it evaluates the trustworthiness of the data, which refers to the instantaneous confidence in the observational data using the data trust module shown in Figure 6.6.

As described in Section 6.3, the gateway node can derive the trust in the sensor data based on the available evidence. Assuming that CPS applications require dense deployment of sensor nodes, the sensor nodes associated with the same gateway node would have correlated observations as a result of the close proximity. Due to this spatial correlation, the observations of neighboring sensor nodes can be used as evidence for the trustworthiness of a sensor observation. This intuition builds on sensor network trust frameworks in [21].

In our trust management framework, reputation of the data source is another element that is used for the trust evaluation of the data. Gateway nodes evaluate and update the reputation of the sensor nodes based on the evidence of other sensor node observations. The reputation of a sensor node, whose observations are supported by the observations of the other sensor nodes, is higher than the reputation of a sensor node, whose observations are not supported by evidence. The relation between the reputation of a data source and the trustworthiness of the data that it provides can be explained as follows. The data source builds reputation based on the long-term behavior and its reputation evolves with time, whereas data trust is instantaneous and it is evaluated for each observation.

Although evidence and reputation elements have been used for trust evaluation of sensor data in the literature, the sensor node's own confidence or uncertainty has not been incorporated for trust evaluation in the existing works. Our trust management architecture takes into account the sensor node's own confidence or uncertainty in its observation to derive the trust in a particular sensor observation. As an example, consider a sensor node that estimates its location based on GPS. There is an associated location uncertainty in GPS-based location estimation due to the received satellite signal strength and the algorithm features used to estimate the location [22]. The sensor node can calculate its position uncertainty estimate based on its GPS module's estimate of error. Our trust management framework feeds the sensor node's own uncertainty estimate into the trust evaluation of the data to account for the possible inaccuracies in its observation.

As shown in Figure 6.6, the data trust module models the trust in an observation $Observation_{ij}$ at the data layer as:

$$Trust_{ij} = f\left(Tsens_{ij},\ Trep_{ij},\ Tconf_{ij}\right)$$

where $Tsens_{ij}$, $Trep_{ij}$, and $Tconf_{ij}$ correspond to the evidence from other sensor node observations, the reputation of the sensor node, and the node's uncertainty in its observation $Observation_{ij}$, respectively. The trust level $Trust_{ij}$ in this observation is calculated with the function f that maps the inputs of the data trust module, that is, $Tsens_{ij}$, $Trep_{ij}$, and $Tconf_{ij}$, to the trust level $Trust_{ij}$. The trust evaluation is instantaneous, and all the terms correspond to the time instance of the observation. Thus, we omit the time (t) notation for simplicity. Note that how we evaluate the trustworthiness of an observation and the trust components (i.e., evidence, reputation, and confidence) depends on the application and the relevant sensor modalities. Intuitively, the function f assigns a higher trust value to an observation if the observation is supported by other observations, the data source has a high reputation score, and its uncertainty in the observation is low.

As described in Section 6.4.1, gateway nodes are responsible for evaluating the trustworthiness of the data collected from sensor nodes and generating a block to record the sensor observations and the trust values on the blockchain. Figure 6.9 shows the gateway node G_i collecting data from the associated sensor nodes $\{S_{i1}, S_{i2}, ..., S_{iK}\}$ through transactions $\{T_{xi1}, T_{xi2}, ..., Tx_{iK}\}$ and generating a new block. Note that every transaction in this block is signed by the sensor node generating the transaction and the gateway node for authentication.

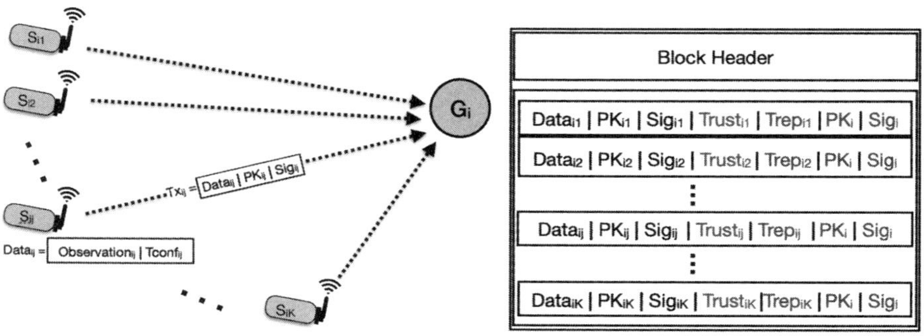

Figure 6.9 The block structure generated by a gateway node.

Next, we will describe how the trust components are determined and the trust value is derived by a gateway node using the simple mapping:

$$Trust_{ij} = Tsens_{ij} \times Trep_{ij} \times Tconf_{ij}$$

- *Confidence of the data source* ($Tconf_{ij}$)*:* When a data source makes an observation, it also estimates the uncertainty in its observation. The confidence of the data source is a measure of how confident the data source is in its observation. We can model the confidence of the data source as a variable $Tconf_{ij} \in [0,1]$, where a higher $Tconf_{ij}$ value corresponds to higher confidence and lower uncertainty in the observation. The data source S_{ij} determines the value of $Tconf_{ij}$ and transmits it to the gateway node G_i together with its observation data $Observation_{ij}$ in the transaction Tx_{ij} given by

$$Tx_{ij} = \left[Observation_{ij} \middle| Tconf_{ij} \middle| PK_{ij} \middle| Sig_{ij} \right]$$

where PK_{ij} and Sig_{ij} are the public key and the signature of the node S_{ij}, respectively. The data source determines its confidence value by using an application-specific confidence model. As an example, consider location data provided by a GPS sensor node. When the received GPS signal is strong, the sensor node estimates lower uncertainty in its location data and would determine a higher confidence value. Conversely, if the received GPS signal is weak, the sensor node would have higher uncertainty in its location estimation and would report a lower confidence value. The sensor node may also report a fixed confidence value if it does not receive any GPS fix.

- *Evidence from other observations* ($Tsens_{ij}$): As discussed in Section 6.3, evidence from multiple observations can be used to establish trust in a sensor observation. In our framework, the gateway node uses multiple observations of correlated data sources to calculate the evidence component for the trust in observations. Assuming that neighboring sensor nodes have correlated observations, the gateway G_i calculates the evidence $Tsens_{ij}$ for the observation $Observation_{ij}$ based on the data received from the set N_{ij} of the neighboring sensor nodes of S_{ij}. During the network initialization, the neighborhood information is recorded in the profiles of the nodes on the blockchain.

While fusing the evidence from multiple data sources, our architecture uses a confidence-weighted evidence fusion method. This allows the gateway node to give more weight to evidences with higher confidence. The gateway node can also use a confidence threshold while fusing the evidences from multiple data sources. An observation $Observation_{ik}$ supporting the observation of the data source $Observation_{ij}$ increases the evidence $Tsens_{ij}$ by a value proportional to its own observation confidence $Tconf_{ik}$, whereas if $Observation_{ik}$ does not support $Observation_{ij}$, it decreases the evidence $Tsens_{ij}$ by a value proportional $Tconf_{ik}$. Based on this intuition, we propose the confidence-weighted evidence fusion method given by:

$$Tsens_{ij} = \frac{1}{\left|\mathrm{N}_{ij}\right|} \times \sum_{S_{ik} \in \mathrm{N}_{ij}} 1_{ik}^{j} \; Tconf_{ik} \tag{6.3}$$

where

$$1_{ik}^{j} = \begin{cases} 1, & \text{if } Observation_{ik} \text{ supports } Observation_{ij} \\ -1, & \text{otherwise} \end{cases}$$

The indicator function 1_{ik}^{j} determines whether the observation from a neighboring data source increases or decreases the evidence value. The condition used in the indicator function is application-specific. As an example, for a Received Signal Strength Indicator (RSSI) based target detection application, the difference between the RSSI values can be compared to a threshold value to determine if the observations support each other or not.

The evidence trust component has a clear dependence on the number of independent observers of a physical phenomenon and the similarity in observations among the observers. However, it is interesting to note the existence of a trade-off between correlations in observations among the nodes, which provides a high trust level, and the additional information value of provided by the observations. Figure 6.10 highlights this trade-off. From a trust perspective, high alignment in observations is desirable, yet new, highly correlated observations add little information to our account of the physical event. Observations with low or no correlation to the majority of observations may be uncorroborated and thus receive low trust levels, yet they have the potential to add significant information value if they are genuine. The green shaded area highlights the sweet-spot operating region for this correlation/information trade-off that delivers fairly well-corroborated observations and that can add meaningful new and potentially less correlated observations.

- *Reputation of the data source* ($Trep_{ij}$): Gateway nodes collect data from the associated data sources over a period of time. During this time period, data sources establish reputations based on their long-term behaviors. The reputation of a data source has an impact on the trust level of its observation data. The observation data provided by a data source with a high reputation is expected to be more trustworthy than the data provided by sources with a low reputation. To take into account the reputations of data sources in the trust evaluation of the data, gateway nodes keep track of the reputations of the associated data sources, which evolve in time.

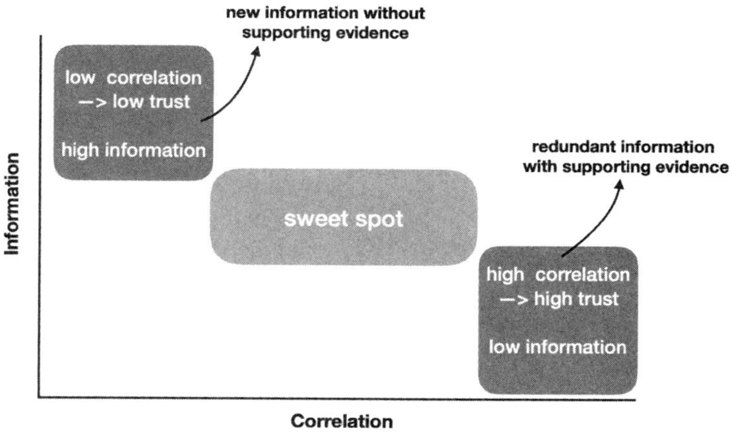

Figure 6.10 The trade-off between information value and correlation. Highly correlated observations increase the trust but provides redundant data (less information).

We propose a reputation update mechanism based on the observation confidence and the evidence of other observations. The update mechanism works by rewarding or penalizing the reputation of the data sources with a value determined by the reported confidences. First, a confidence threshold $conf_{sh}$ is used to determine whether the data source S_{ij} has high or low confidence in its observation. If the confidence of the data source is high ($Tconf_{ij} \geq conf_{th}$), a high reputation update value (ΔRep_H) is chosen by the gateway node. Conversely, if the confidence of the data source is low ($Tconf_{ij} < conf_{th}$), a lower reputation update value (ΔRep_L) is chosen, where $\Delta Rep_L < \Delta Rep_H$. Then the gateway node uses an evidence threshold ($evidence_{th}$) to decide if the observation is supported by other observations or not. If the observation is substantiated by other nodes ($Tsens_{ij} \geq evidence_{th}$), the reputation of the data source is increased by the reputation update value determined by the gateway before. Conversely, if the observation is refuted by other nodes ($Tsens_{ij} < evidence_{th}$), the reputation of the data source is decreased by the same reputation update value.

Note that malicious data sources may use the reputation update mechanism to report tampered observation data with low confidence. In this case, the reputation of the malicious data sources would not suffer a significant drop for each observation due to the low confidence reported. This security attack by the malicious data sources can be addressed by:

- *Confidence-weighted evidence fusion*: The design of the function weighting high uncertainty measurements in the trust level calculation should ensure that the impact of the high uncertainty measurements on the trust level of the observation data would be low.
- *Penalizing repetitive low confidence observations*: The reputation score update mechanism should penalize repetitive low confidence observations from the same node.

As discussed before, the components of the data trust module and the trust evaluation function are application-specific. Thus, the data trust block in our architecture is modular and can be adapted to include different trust components and trust calculation functions for other application scenarios. In the data trust module, the spatial correlation among neighboring sensor nodes is used to calculate the evidence. For some other CPS applications, the temporal correlation of observations or multimodal sensor fusion mechanisms can be used to derive the evidence for observations depending on the spatio-temporal properties of the physical phenomenon being observed and the sensing modalities of the data sources.

For each sensor observation $Observation_{ij}$, the associated gateway node G_i calculates the reputation $Trep_{ij}$ of the data source S_{ij} and the evidence from other data sources $Tsens_{ij}$. The data source reputation values and the observation trust values derived from the trust components are included as part of the transaction that records the occurrence of the observation in the block generated by the gateway node as presented by Figure 6.9. Once this block is validated by the other gateway nodes maintaining the blockchain network, it is appended to the blockchain and provides an immutable and auditable record of the observations, associated trust values of the observations, and the reputation of the data sources in the blockchain.

6.4.2.2 Reputation of Gateway Nodes

In Section 6.4.2.1, we described the details of the data trust module and how gateway nodes evaluate the trustworthiness of data collected from associated data sources. The data trust evaluation is incorporated into the transactions to generate new blocks for the blockchain layer. Next, we will present the details of the reputation module, which is responsible for updating the reputations of the gateway nodes. The gateway reputations establish entity-centric trust relationships among the gateway nodes, which is used by our adaptive block validation mechanism to integrate the data trust mechanism to the blockchain layer.

Recall that, in our architecture, the overlay blockchain network is maintained by the gateway nodes. Thus, when a gateway node generates a new block as shown in Figure 6.9, this block is shared with the other gateway nodes for block validation before being appended to the blockchain.

The block validation mechanism used in our trust architecture involves two steps:

- *Validating the data transactions*: The validator nodes check the public keys of the data sources and their signatures of the transactions included in the block.
- *Validating the data trust values*: The validator nodes recalculate the trust values of the observations assigned by the gateway node with the data available in the generated block and on the blockchain.

Inspired by the reputation model in Chapter 4, the gateway reputation module shown in Figure 6.7 tracks the long-term behavior of gateway nodes and adapts the block validation depending on the reputation of the current gateway node. To provide end-to-end trust for CPS applications, our reputation module further integrates the data trust mechanism to the block validation process by considering three components:

- $T(G_i)$: How much other validator nodes trust G_i based on G_i's trust value assignment to the observations;
- $B(G_i)$: How much other validator nodes trust G_i based on G_i's block mining;
- $Ext(G_i)$: The external sources of G_i's reputation $Ext(G_i)$ from other systems that can be fed into the gateway node's reputation score.

We can model the reputation score of the gateway node G_i, $Rep(G_i) \in [Rep_{min}, Rep_{max}]$, as:

$$Rep(G_i) = g\left(T(G_i), B(G_i), Ext(G_i)\right)$$

where g is a function mapping the reputation components to the reputation value of the gateway.

Every time the gateway node generates a new block, its reputation is updated by the validator nodes based on the validity of sensor transactions and the correctness of the associated trust values. Intuitively, the reputation of the gateway node increases if the generated block is validated and decreases otherwise. The reputation increases and reduction steps can be chosen such that it is harder for the gateway node to build a reputation than to lose it.

6.4.3 Lightweight Blockchain Architecture

Inspired by the Lightweight Scalable Blockchain (LSB) that is optimized for CPS requirements as described in Chapter 4, we propose a private blockchain for our trust architecture, which is built on a lightweight block generation mechanism, reputation-based adaptive block validation, and distributed consensus among blockchain nodes.

6.4.3.1 Lightweight Block Generation Mechanism

Our end-to-end trust architecture for CPS applications is built on the layered structure shown in Figure 6.5. The data layer consists of the sensor nodes that collect observation data from the physical domain and send the collected data to the gateway nodes, whereas the blockchain layer consists of the gateway nodes that participate in block generation, block validation, and distributed consensus in a private blockchain network. Contrary to the public blockchains, where nodes do not need permissions to participate in the blockchain network, in private blockchains, nodes need to have permissions. During the network initialization phase, the gateway nodes register to the blockchain network. Since the gateway nodes are registered to the network and they have permissions to generate blocks, they do not need to compete for block generation using computationally expensive block mining mechanisms, such as POW. Instead, our lightweight block generation mechanism assigns periodic intervals for gateway nodes to generate new blocks. Gateway nodes wait for their turns to multicast their blocks to other gateway nodes for validation. The assignment of the time intervals for block generation can be adjusted based on the application requirements, the sensor data rate, and the latency of data collection and block generation.

6.4.3.2 Adaptive Block Validation

Block validation in our architecture involves both validating the data transactions and trust value assignments in a block, making it computationally demanding. Thus, we propose a block validation mechanism that adapts the block validation scheme based on the reputation of the block generating node $Rep(G_i)$ and the number of validator nodes N_{val}. The introduced adaptive block validation mechanism is managed by the gateway reputation module of our architecture. In this section, we will describe how integration of trust management improves the block validation.

- *The effect of the reputation of the block generating node $Rep(G_i)$:* In the existing blockchains, every validator node validates all the transactions in a new block before appending it to its blockchain. However, in our adaptive block validation scheme, each validator node randomly validates a percentage of the transactions in the block depending on the reputation of the block-generating node. We can explain the intuition behind using reputation for adaptive block validation by considering the attack success of a malicious block-generating node:

$$P(\textit{successful attack}) = P(\textit{attack succeeds} \mid \textit{attack})\ P(\textit{attack})$$

Assume that we want the probability of a successful attack by a block generating node to be lower than a *P*(*successful attack*) threshold. If the node has high reputation, we expect that its attack probability *P*(*attack*) would be low. Then, for a target *P*(*successful attack*), the system can tolerate a higher conditional successful attack probability *P*(*successful attack*| *attack*). In terms of block validation, that corresponds to validating a smaller number of transactions in a block generated by a gateway with high reputation. We use this intuition to model the relative effect of the reputation on the percentage of transactions to be validated with a linearly decreasing function. If $Rep(G_i)$ is high, validator nodes validate a small percentage of the transactions. If the block is validated, $Rep(G_i)$ is increased, which means that the next time G_i generates a new block, validator nodes will need to validate a smaller percentage of the transactions. However, if the block cannot be validated, $Rep(G_i)$ is decreased. In this case, the next time G_i generates a new block, validator nodes need to validate a larger percentage of the transactions in the block.

- *The effect of the number of validators* N_{val}*:* Our block validation mechanism also adapts the percentage of the transactions to be validated depending on the number of validators N_{val}. For a target *P(successful attack)* threshold, as the number of validators increases, the percentage of transactions required to be validated randomly by each validator node decreases.

  Assume that there are Tx_{total} transactions in the block generated by a malicious gateway node, and Tx_{inval} of these transactions are invalid (i.e., they do not have the correct signatures), or the assigned trust values are misrepresented. Following our adaptive block validation logic, validator nodes validate only a percentage of the transactions in the generated block. Consequently, there is a risk of not detecting any invalid transactions in a given block. The probability of not detecting any invalid transactions by N_{val} validator nodes is given by:

$$P(block\ validated\ |\ there\ are\ Tx_{inval}\ invalid\ transactions) = \left(\frac{\binom{Tx_{total} - Tx_{inval}}{Tx_{val}}}{\binom{Tx_{total}}{Tx_{val}}} \right)^{N_{val}} \tag{6.4}$$

  where Tx_{val} is the number of transactions to be validated randomly by each validator node. There are $\binom{Tx_{total}}{Tx_{val}}$ ways to choose a subset of Tx_{val} transactions to be validated out of which $\binom{Tx_{total} - Tx_{inval}}{Tx_{val}}$ does not include any invalid transactions.

Figure 6.11 shows the probability of not detecting any invalid transactions in a block (i.e., successful attack probability) as a function of the number of transactions validated randomly by each validator node for $N_{val} = 1, 5, 20$ validator nodes when $Tx_{total} = 100$. As expected, when we increase the number of validator nodes, we can reduce the percentage of transactions that need to be validated without sacrificing the invalid transaction detection probability. For instance, when $N_{val} = 20$, the successful attack probability is below 0.08% when

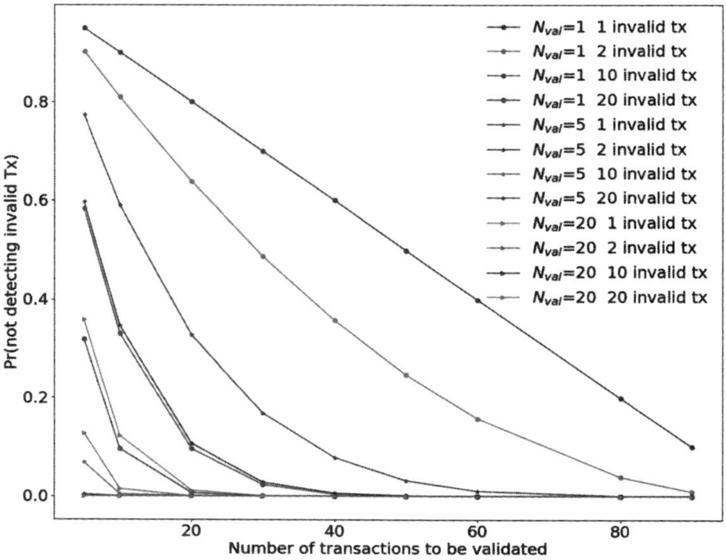

Figure 6.11 Successful attack probability as a function of the number of the transactions to be validated for $Tx_{total} = 100$ and $N_{val} = 1, 5, 20$.

each validator validates only 30% of the transactions. If there is more than one invalid transaction in the block, the probability of a successful attack decreases significantly.

In Figure 6.12, we also show the minimum number of validators required to achieve a target successful attack probability threshold. As an example, when $N_{val} = 25$ and $Tx_{val} = 25$ for $Tx_{total} = 100$ and $Tx_{inval} = 1$, the successful attack probability becomes less than 0.1%. The same threshold can also be achieved by $N_{val} = 14$ and $Tx_{val} = 40$.

Considering the effects of the reputation of the block-generating node and the number of validator nodes on the block validation performance, our architecture proposes an adaptive block validation mechanism, where each validator node validates randomly only a portion of the transactions in the block (i.e., the percentage of the validated transactions (PVT)). Based on our observations, we empirically model the PVT as a function of the reputation of the block generating node (Rep) and the number of validator nodes (N_{val}) as:

$$PVT\left(Rep, N_{val}\right) = \left(\gamma_0 + \gamma_1 \ \ Rep\right) \times e^{-\ddot{a} N_{val}} \times 100\% \tag{6.5}$$

where Δ is a controlling parameter determining the effect of N_{val} on PVT, and γ_0 and γ_1 are the parameters of an affine function determining the effect of

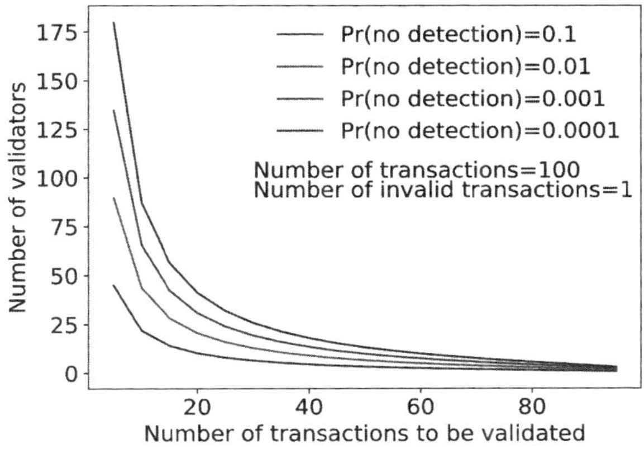

Figure 6.12 Number of validators required for a target probability threshold of not detecting an invalid transaction in a block as a function of number of validated transactions by each validator.

Rep on *PVT*. In [18], we showed that the adaptive block validation scheme can reduce the computational cost of block validation process significantly and improve the scalability and latency of the trust architecture.

6.4.3.3 Distributed Consensus Mechanism

Since our adaptive block validation mechanism requires the validators to randomly choose and validate a number of transactions in a block rather than all the transactions, there is a need to share the result of the validations among the validator nodes. Thus, after randomly validating the transactions in a block, each validator multicasts the result of the validation. If a validator node cannot find any invalid transactions among the transactions that it validates, it multicasts a "VALID" message to the other validator nodes. Otherwise, if the validator node finds any invalid transactions in the block, it multicasts the transaction ID of the invalid transactions with an "INVALID TRANSACTION ID" message to notify the validator nodes about the invalid transactions in the block. The other validator nodes receiving the "INVALID TRANSACTION ID" message also tries to validate the transactions identified by the message. If they find at least one invalid transaction in the block, the block is rejected. Otherwise, if all transactions are verified to be valid, the block is appended to the blockchain. Figure 6.13 summarizes how the observation data collected by the sensor nodes and the trust evaluation of the data assigned by the gateway nodes are stored on an immutable and auditable blockchain in our end-to-end trust architecture framework.

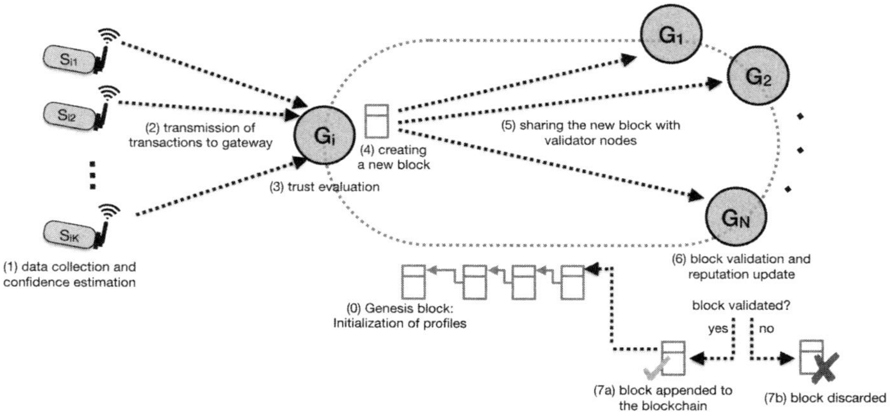

Figure 6.13 The end-to-end trust framework for recording sensor observations and the trust evaluation of the observations on an immutable and auditable blockchain for CPS applications.

6.4.4 Case Study

In the previous sections, we introduced the architecture and the components of our blockchain-based, end-to-end trust management framework. The data trust and gateway reputation modules of our trust framework are modular and can be customized for different CPS applications, making the framework applicable for a diverse range of use cases, such as smart cities, supply chains, and industrial and defense applications. In this section, we apply our trust framework to an indoor target localization application in a smart factory environment, where IoT devices collect data from the factory to monitor all stages of the production.

In this industrial IoT setting, sensor nodes, human operators, and autonomous machines and vehicles send their data to edge devices acting as the gateway nodes. Before this data is recorded on the blockchain, its trustworthiness is evaluated by the associated gateway node using the data trust module. Consider a scenario where an automated vehicle enters a restricted production area due to a malfunctioning proximity sensor. The vehicle sends its inaccurate location data with high confidence to an edge device. Once the edge device receives the data, it collects information from other sources in the vicinity of the vehicle, such as other automated vehicles and machines, sensor nodes, and human operator reports, and calculates the weights corresponding to the information collected from these sources. As a result, the measurements with higher confidence obtained recently from the sources in the vicinity of the vehicle have a higher effect on the verification of the vehicle's location data. By considering the weighted information from the other sources, historical reputation and the confidence of the vehicle's location data, the edge device can arrive at an informed assessment of the trustworthiness of the data sent by the vehicle. Next, the edge device generates a new block that stores the data collected from the IoT devices and other data sources, and the trust evaluations for the collected data.

At the blockchain layer, the edge device shares the generated block with other edge devices for validation. Using the gateway reputation module, and the adaptive block validation mechanism, the new block is either appended to the blockchain or discarded based on the validation result. Our end-to-end trust framework for blockchain-based CPS applications improves the trust of data both at the origin and on the blockchain record.

6.5 Security Analysis of the End-to-End Trust Framework

For the security analysis of the blockchain based end-to-end trust architecture for CPS applications, we consider sensor nodes, gateway nodes, and external agents as potential attackers. Each node is identified by a public key and has a secure mechanism to generate and keep the private key. The architecture uses a private blockchain, where sensor and gateway nodes register to the network with their public keys during network initialization by a trusted agent. Based on these assumptions, we will analyze potential attack scenarios by different agents in the system:

- *Malicious sensor nodes*: In CPS applications, the sensor devices that collect data from the physical domain are prone to security attacks. In many cases, these sensor devices can be tampered with easily to produce inaccurate observations. Based on the evidence of other node observations, the reputation of the sensor nodes and the reported confidence levels, our data trust module assigns trust values to sensor observations. The tampered observation will be assigned a trust value as long as there are enough numbers of honest sensor nodes with high reputations and confidence levels to refute the tampered observation. Besides, the data trust module decreases the reputation of the malicious sensor node for future observations. Malicious sensor nodes can also collude to increase the successful attack probability by reporting tampered observations that support each other, exploiting the evidence component of the data trust evaluation. However, the collusion attack may require a large number of malicious sensor nodes with high confidences in order to be effective.
- *Malicious gateway nodes:* In our architecture, gateway nodes are responsible for both block generation and block validation in the overlay blockchain network. A block-generating malicious gateway node can try to tamper with the sensor observations or the trust values assigned to these observations. However, during block validation, validator nodes can de-

tect data tampering by verifying the signatures of the data sources and can also detect inaccurate trust value assignments by recomputing the trust values. After the validators detect invalid transactions in a block, the block is rejected by the network. Furthermore, the reputation of the malicious block-generating node is downgraded, which would make a successful attack even harder in the future due to a stricter validation process imposed by the lower reputation score. Malicious gateway nodes, which repeatedly attempt to create invalid blocks, can be isolated from the blockchain network and the data sources connected to that node are associated with a new gateway node.

A block-generating node may also attempt an on-off attack by building a reputation to use it occasionally for malicious activities and not to get detected due to its high reputation value. To mitigate this attack, the adaptive block validation mechanism may determine the number of transactions to be validated as a normal random variable, whose mean value is chosen based on the reputation of the block generating node. Due to this randomization, even blocks generated by block-generating nodes with high reputations can be subject to strict validation of transactions.

A malicious validator node may try to waste the network resources, forcing all the transactions in the block to be validated by all validator nodes by broadcasting "INVALID TRANSACTION ID" messages for all transactions. To mitigate this attack, during the consensus period, each validator is allowed to multicast only one message, either confirming the valid block or containing the transaction ID for only one invalid transaction. Malicious validator nodes can also collude to validate invalid blocks. The success probability of this attack decreases with the number of validator nodes. When the blockchain network has a low number of validator nodes, the choice of block-validating nodes can be randomized to mitigate the collusion of malicious block validating nodes.

- *Malicious external nodes:* An external agent may try to attack the network by impersonating a sensor node or a blockchain node. Note that, since all transactions are signed using the private keys of the nodes, whose public keys are known and used for verification of the transactions, this attack requires the external agent to have access to the private key of the attacked node.

6.6 Conclusions

This chapter focused on trust management for blockchain-based CPS applications. We started by analyzing the trust problem for CPS applications, which requires establishing trust in data, interaction, and application layers, and the interactions between these layers. Then we studied the existing data-centric and entity-centric approaches for trust management. Recently, blockchain has been proposed as a promising technology for establishing trust for CPS applications. However, blockchain mechanisms alone cannot guarantee end-to-end trust for CPS applications. Thus, we developed a multilayered trust management framework to establish end-to-end trust that can be applied to a diverse range of blockchain-based CPS applications. The framework uses a data trust module to evaluate the trustworthiness of observations at the data layer by using fusion of information from different data sources, confidence of observations, and reputation level of data sources. At the blockchain layer, a reputation module is used to establish entity-centric trust relations among the gateway nodes by tracking and updating the reputations of the gateway nodes. The gateway reputations are used by our adaptive block validation mechanism to integrate the data trust mechanism to the blockchain layer. The end-to-end trust framework is built on a private blockchain architecture with lightweight block generation, adaptive block validation, and distributed consensus mechanisms. Finally, we presented the security analysis of the trust management framework considering the common security attack scenarios.

References

[1] Wang, J., et al., "Distributed Trust Management Mechanism for the Internet of Things," *Applied Mechanics and Materials*, 2013, pp. 347–350.

[2] Vukolic, M., "The Quest for Scalable Blockchain Fabric: Proof-of-Work vs. BFT Replication," *International Workshop on Open Problems in Network Security*, 2015, pp. 112–125.

[3] Wood, G., "Ethereum: A Secure Decentralised Generalised Transaction Ledger," Ethereum Project Yellow Paper, Vol. 151, 2014.

[4] Intel: Sawtooth Lake, 2017. https://intelledger.github.io/.

[5] Babich, V., and G. Hilary, "Distributed Ledgers and Operations: What Operations Management Researchers Should Know About Blockchain Technology," *SSRN Electronic Journal*, 2018.

[6] Jayasinghe, U., et al., "Data Centric Trust Evaluation and Prediction Framework for IoT," *2017 ITU Kaleidoscope: Challenges for a Data-Driven Society (ITU K)*, 2017, pp. 1–7.

[7] Raya, M., et al. "On Data-Centric Trust Establishment in Ephemeral Ad Hoc Networks," *IEEE INFOCOM 2008-The 27th Conference on Computer Communications*, 2008, pp. 1238–1246.

[8] Hall, D. L., and J. Llinas, "An Introduction to Multisensor Data Fusion," *Proceedings of the IEEE,* Vol. 85, No. 1, January 1997, pp. 6–23.

[9] Shafer, G., *A Mathematical Theory of Evidence,* Princeton, NJ: Princeton University Press, 1976.

[10] Sicari, S., et al., "Security, Privacy and Trust in Internet of Things: The Road Ahead," *Computer Networks,* Vol. 76, 2015, pp 146–164.

[11] Yan, Z., P. Zhang, and A. V. Vasilakos, "A Survey on Trust Management for Internet of Things," *J. Netw. Comput. Appl.,* Vol. 42, June 2014, pp. 120–134.

[12] Lu, Z., et al., "A Privacy-Preserving Trust Model Based on Blockchain for VANETs," *IEEE Access,* Vol. 6, 2018, pp. 45655–45664.

[13] Kang, J., et al., "Blockchain for Secure and Efficient Data Sharing in Vehicular Edge Computing and Networks," *IEEE Internet of Things Journal,* 2018.

[14] Kchaou, A., R. Abassi, and S. Guemara, "Toward a Distributed Trust Management Scheme for Vanet," *Proceedings of the 13th International Conference on Availability, Reliability and Security,* 2018, pp. 53:1–53:6.

[15] Kang, J., et al., "Toward Secure Blockchain-Enabled Internet of Vehicles: Optimizing Consensus Management Using Reputation and Contract Theory," *IEEE Transactions on Vehicular Technology,* Vol. 68, No. 3, March 2019, pp. 2906–2920.

[16] Dorri, A., et al., "LSB: A Lightweight Scalable Blockchain for IoT Security and Anonymity," *Journal of Parallel and Distributed Computing,* Vol. 134, 2019, pp. 180–197.

[17] Bahri, L., and S. Girdzijauskas, "Trust Mends Blockchains: Living up to Expectations." *2019 IEEE 39th International Conference on Distributed Computing Systems (ICDCS),* 2019.

[18] Dedeoglu, V., et al., "A Trust Architecture for Blockchain in IoT," *16th EAI International Conference on Mobile and Ubiquitous Systems: Computing, Networking and Services (MobiQuitous),* Houston, TX, November 12–14, 2019.

[19] Wang, W., K. Lee, and D. Murray, "Building a Generic Architecture for the Internet of Things," *2013 IEEE Eighth ISSNIP,* Melbourne, VIC, 2013, pp. 333–338.

[20] Pattem, S., B. Krishnamachari, and R. Govindan, "The Impact of Spatial Correlation on Routing with Compression in Wireless Sensor Networks," *ACM Transactions on Sensor Networks (TOSN),* Vol. 4, No. 4, 2008, p. 24.

[21] Han, G., et al., "Management and Applications of Trust in Wireless Sensor Networks: A Survey," *Journal of Computer and System Sciences,* Vol. 80, No. 3, 2014, pp. 602–617.

[22] Serr, K., T. Windholz, and K. Weber, "Comparing GPS Receivers: A Field Study," *URISA Journal,* Vol. 18.2, 2006, pp. 19–23.

7

User Anonymity in Blockchain

7.1 Introduction

One of the key features of the blockchain is the anonymity of the transaction generators (see Chapter 3). Blockchain participants employ pseudonyms, which is typically their public key (PK^+) that is included in transactions. This hides the real-world identity of the user, as shown in Figure 7.1, and thus provides anonymity. The user can prove his identity by signing a transaction with his private key (PK^-), which corresponds to the PK^+ included in the transaction. This enables the anonymous verification of the user. The anonymity offered by the blockchain increases the privacy of the users. Anonymity refers to hiding the identity of the user while privacy pertains to hiding the content of a transaction (i.e., the exchanged message) [1]. Malicious entities may attempt to deanonymize blockchain users using various methods, for example, classifying transactions or using off-chain information (see Section 7.2.3 for a detailed discussion).

Recall from Chapter 3 that blockchain was first introduced in Bitcoin [2], the first decentralized cryptocurrency. Various works have analyzed the anonymity of users in cryptocurrencies. The authors in [3] outlined eight attributes that capture various properties of anonymity in cryptocurrencies:

1. *Anonymity of an entity* is the ability of a malicious node in the network to identify an anonymous node within a group of nodes, known as an anonymity set. A larger anonymity set suggests higher anonymity for the users.

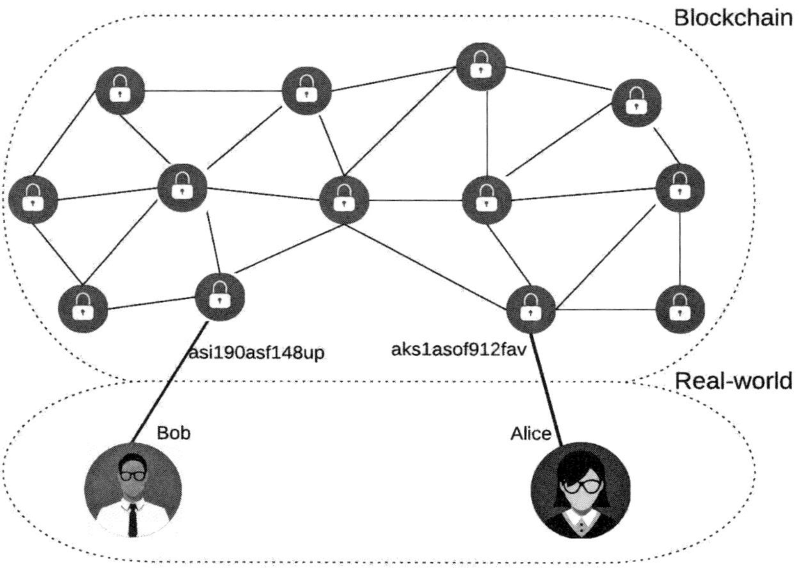

Figure 7.1 Anonymity in blockchain.

2. *Recipient anonymity* is similar to the entity anonymity but for the outputs of the transaction. This attribute refers to the ability of the malicious nodes to link a message output to a particular node in the network.
3. *Unlinkability* refers to the impossibility of linking multiple transactions to a single node.
4. *Untraceability* refers to the impossibility of linking any transaction in the network to its generator.
5. *Fungibility* refers to the equality of the currency assets in a blockchain (i.e., no asset should be blacklisted based on the previous history of the node's interactions).
6. *Hidden transaction values* refer to the impossibility of identifying the exact values as input and output to a transaction. The malicious nodes can identify the flow of transactions by monitoring the transaction values and thus cluster the transactions to deanonymize the users.
7. *Metadata unlinkability* refers to the impossibility of linking available metadata information (e.g., Internet Protocol (IP) address) to a transaction. If malicious nodes can link the metadata to the transactions, they can identify all transactions generated by the same node, which compromises the node's privacy and anonymity.

8. *Deniability* refers to the ability of a malicious node to deny participating in the generation of a particular transaction.

As outlined in Chapter 3, blockchain applications are not limited to cryptocurrencies and a broad range of applications have been defined in recent years including CPS. In CPS applications, the transactions include data or are linked to off-chain data. In addition to users, a broad range of CPS devices also generate transactions to exchange information and offer personalized automated services to the users. In contrast, this introduces new threats to user anonymity and privacy. The authors in [4] outlined that, to protect the privacy of the users, blockchain systems must satisfy two requirements:

1. *Identity privacy:* This refers to unlinkability of the transactions with the real identity of the user. Accordingly, both the sender and the recipient must remain anonymous to prevent malicious nodes from linking transactions and deanonymizing a user.
2. *Transaction privacy:* This refers to protecting the content of the transaction (i.e., the user data) from being revealed to the participating nodes in the public blockchain. Malicious nodes may attempt to deanonymize a user based on the data stored in the transaction (e.g., the location of a smart home).

In this chapter, we study the issue of anonymity in the blockchain. Most of the existing anonymity studies focus on the anonymity of the users in cryptocurrencies. Thus, we also consider cryptocurrency applications of the blockchain in our discussions. However, the attacks and methods for enhancing anonymity are generic and can be extended to CPS applications. In the rest of this chapter, we first study threats against user anonymity in Section 7.2. We then discuss the frameworks to protect the anonymity of the users in Section 7.3. To enhance their anonymity, a user may employ multiple PK^+ as their identity which raises the issue of managing multiple keys. We study this challenge and the existing solutions in Section 7.4. In Section 7.5, we study the anonymity of the users in CPS applications, and finally Section 7.6 concludes the chapter.

7.2 Threats Against User Anonymity

Blockchain employs PK^+ as pseudonyms to identify the users, which introduces a level of anonymity. The malicious nodes may attempt to deanonymize the users by analyzing the blockchain transactions, linking off-chain data to transactions, and analyzing network-layer information. This section provides a review on the existing deanonymization methods. We categorize the existing works in four categories: (1) active interactions (Section 7.2.1), (2) analyzing network traffic (Section 7.2.2), (3) analyzing transactions (Section 7.2.3), and (4) analyzing off-the-chain information (Section 7.2.4). Table 7.1 provides an overview of these methods.

7.2.1 Active Interaction

The first deanonymization method is active interaction. In this approach, the attacker identifies the nodes by directly interacting with them through transactions. As an example, an attacker may attempt to buy goods from an anonymous user assuming that the user is selling goods. To pay the price of a certain item, the buyer must know the PK^+ of the seller, which enables the buyer to start gathering information about the seller. As another example, the attacker may deposit money and execute a withdrawal at a coin exchange service (e.g., Mt. Gox [5]). The authors in [6] attempted to deanonymize the users in Bitcoin by conducting active interactions. The authors in [6] interact directly with 31 service providers including mining pools, wallet services, bank exchanges, non-bank exchanges, vendors, and gambling sites through 344 transactions. Upon receipt of the transactions from the service provider (i.e., receiver), the associated PK^+ is tagged as an address owned by a particular receiver. To maximize the number of PK^+ collected for each service provider, the sender sends multiple requests and captures different PK^+ employed by the receiver. In some cases, the service provider provides additional anonymity measures to protect against malicious activities, which makes it impossible to obtain their PK^+. As an example, the Eligius mining pool pays the mining reward to the accounts of the users immediately after the block is mined. Thus, the reward transaction does not contain the PK^+ of Eligius. To increase the number of identified nodes, the authors also rely on the information available from off-chain sources. For example, a user may reveal his or her PK^+ in a forum or the PK^+ of a company may be publicly available on the company's website.

Table 7.1
A Summary of the Threats Against User Anonymity

Threat	Requirements
Active interaction	Demands direct interaction with the target; demands knowledge of the PK^+ of the target; collects network-layer information of the user
Analyzing network traffic	Demands real-time access to the network; demands eavesdropping on network traffic; collects network-layer information of the user and links with keys; clusters keys based on IP and location
Analyzing off-the-chain data	Demands access to external resources where users share data; links off-chain data with transactions
Analyzing transactions	Demands access to the history of transactions in blockchain; classifies transactions or generating nodes

To classify the transactions, two heuristics are employed that are discussed below:

- *Heuristic 1: If two (or more) addresses are inputs of a single transaction, they belong to the same user.*
- *Heuristic 2: If a transaction has one input and one output, then both input and output belong to the same user. Such transaction is known as one-time change address transaction and is controlled by the same user.*

To study the outlined heuristics, two graph structures were constructed. The first was the transaction graph, where transactions are represented as vertices of the graph. A directed edge between transaction t_1 and transaction t_2 indicates that the input of t_1 is used as the output of t_2. The second was the PK^+ graph, where PK^+ are represented as vertices of the graph. Similar to transaction graph, the input and output flow is represented as edges of the graph. Using the first heuristic, the network was clustered into 5,579,176 clusters of users (the network initially contained 16,086,073 transactions). Some of these clusters can be linked to a particular user; for example, 20 clusters belong to the Mt. Gox mining pool. To further increase the accuracy of the classification, the authors employed the second heuristic, which covers shadow addresses where a single user changes the PK^+ of his or her transaction. To do so, the users pay the coin to their new address. The second heuristic produced 3,384,179 distinct clusters where the identity of 2,197 users was revealed.

In summary, using active interaction methods the identity of the users can be classified. This method is applicable only in cases where the target node is providing some services in the blockchain. However, most users do not offer services or sell goods. In the next section, we study passive deanonymization by analyzing network traffic.

7.2.2 Analyzing Network Traffic

The blockchain runs on top of a peer-to-peer network where the participating nodes broadcast transactions using IP addresses (see Chapter 3). By sniffing the real-time traffic in the underlying peer-to-peer network, one can extract the IP address associated with anonymous transactions, which empowers the malicious nodes to collate multiple transactions generated by the same node. However, note that the network-layer information is not stored in the blockchain database. The blockchain users may employ Tor [7] to hide their real IP address from the network and thus protect against attacks based on network traffic analysis.

In the underlying peer-to-peer network, the first node that initiates the transmission of a transaction can be assumed as the transaction generator given that it is the only node that is aware of this transaction [8]. The underlying protocol that is used to broadcast transactions in the blockchain can have a significant impact on the ability of malicious entities attempting to classify and identify the users. Initially, Bitcoin employed a gossiping protocol to flood transactions. The gossiping protocol distributes a message within a certain round. Initially, the generator sends the message to a random set of nodes. In the next round, the recipient nodes send the message to another set of random nodes. This process continues until all the participants receive the message. Multiple studies have demonstrated that the gossiping algorithm compromises the privacy of the users. Consequently, in 2015, Bitcoin changed the underlying algorithm to diffusion spreading [9, 10]. In the latter, each node transmits the message to one of its neighbors who has not received it with an exponential independent delay. The node starts the exponential clock once it receives a message. The authors in [11] studied the anonymity of both gossiping algorithm and the diffusion spreading by eavesdropping on the underlying peer-to-peer network to detect the first nodes that initiated a transaction (also known as first relayer). They employed machine learning with two estimators: first timestamp estimator and centrality-based estimator. In the former, the adversary monitors the infection time of each node and attempts to find the first timestamp (i.e., the transaction generator). In the centrality-based method, the estimator creates the network graph based on the timestamp of the transactions. For each potential source node (say, N_s), the estimator counts the total number of messages that it received from nodes in the adjacent subtrees of N_s. The node with equal number of messages in all subtrees is selected as the message generator. The results demonstrate that the diffusion algorithm slightly improves anonymity compared with the gossiping algorithm. However, still both methods offer poor anonymity to the participants.

The authors in [12] argued that the blockchain users can be deanonymized even if they employ Tor. The study suggested that the users that are connecting through the same ISP can also be identified even if they employ the same public IP address. To deanonymize the users, the malicious node first identifies the entry points (i.e., the nodes used by the users to connect to the blockchain). The malicious node connects to as much entry points as possible to ensure that it receives information about new nodes joining the network and records the IP of the new nodes and the IP of the associated entry point. The malicious node monitors transactions generated by each node and links those to their IP addresses and thus deanonymizes the users. Using this method, multiple PK^+ of a user used in a single session can be linked back to the user even if the PK^+ do not belong to the same ledger. With less than 50 connections per entry points, the malicious node succeeds in identifying the IP address of 11% of the participating nodes in the Bitcoin network. By increasing the false-positive tolerance level, the malicious node can identify 35% of the participants.

7.2.3 Analyzing Transactions

The users interact and exchange coins through transactions, thus, the transactions potentially reveal the flow of coin between the users and analyzing these patterns of transactions may potentially lead to user deanonymization.

To analyze the blockchain transactions, the malicious node must first represent the flow of transactions as a graph that is used to classify users. The main steps in preprocessing transactions are as follows [13]:

- *Transaction graph:* Blockchain can be represented as an acyclic graph $G = \{T, E\}$, where T is the set of transactions and E is the set of edges that connect transactions and represents the flow of input and output. Figure 7.2(a) shows an example of a transaction graph.
- *Address graph:* By analyzing the transaction graph, one can construct the address graph $G_a = \{P, E`\}$, where $P$ is the set of Bitcoin addresses and $E`$ is the set of edges connecting the nodes. Figure 7.2(b) shows the address graph.
- *User/entity graph:* By analyzing the address graph along with some heuristics (discussed later in this section), one can construct a user graph as $G_u = \{U, E``\}$ where $U$ is the set of PK$^+$ that potentially may belong to the same user and $E``$ is the edge connecting the users.

In cryptocurrencies, the input of a transaction (say, T_A) is a reference to the output of another transaction (say, T_B). This enables the owner of the output of T_B to spend coins. The node that initiates T_A must prove that it owns the output of T_B that is going to be spent, which is done by signing T_A using the

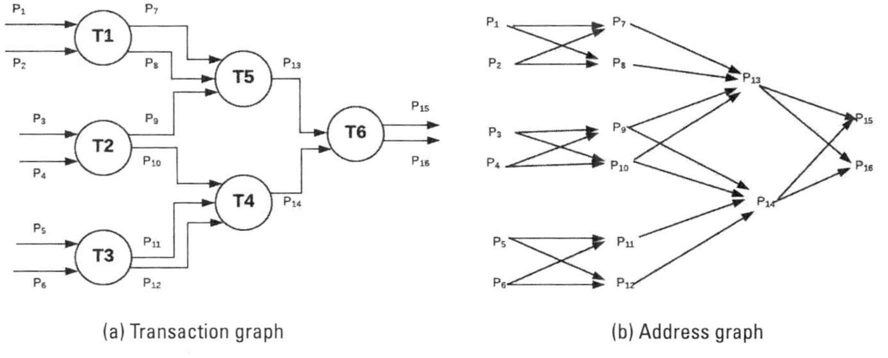

(a) Transaction graph (b) Address graph

Figure 7.2 The structure of (a) the transaction graph and (b) the corresponding address graph.

PK^- corresponding to the PK^+ in the output of T_B. This creates a link between T_A and T_B as the output of T_B is owned by the same user who generated T_A. A transaction may have multiple inputs; thus, it is fair to consider that all outputs belong to the same user.

The authors in [14] employed shadow addresses and multi-input transactions as inputs as heuristics to construct user graphs and subsequently deanonymize the users. A simulator was implemented that mimics the behavior of the users in the real world. For each payment, the user made in the real world, the simulator generates a transaction in Bitcoin with the same timestamp and price. The shopping behavior of the staff and students in a university is fed to the simulator. The results of the analysis showed that the profile of 40% of the staff could be constructed with 80% accuracy even if the users manually transferred their coins among different addresses to enhance their anonymity level.

The authors in [15] studied the impact of mapping the time pattern in which transactions are generated with the geographical locations to identify users. They downloaded the Bitcoin blockchain database, which contains 38,886,789 PK^+. The authors then analyzed the input and output of the transactions and tagged the transactions that use the input of a single transaction or multiple transactions as belonging to the same person. This allowed them to distinguish 17,472,156 owners in the blockchain. Next, the authors attempted to identify the location of the users using the public information available on the Internet (e.g., some companies reveal their location and their PK^+ on their website). There were 1,800 addresses found with known locations. To focus the study on the impact of the time zone on identifying the users, only owners in different time zones were selected, resulting in a total of 518 owners. The shopping behavior of the nodes was then analyzed, and they were able to accurately identify 72% of the owners spread across 11 UTC time zones.

7.2.4 Analyzing Off-the-Chain Information

The blockchain users may reveal information in publicly available databases that potentially may either disclose or assist the disclosure of their real identity. As an example, users may reveal their PK^+ in a forum, or a company may reveal its PK^+ to its users or on the webpage of the company. Malicious nodes may employ such external information to identify the real-world ID of an anonymous user.

The authors in [16] studied the impact of cookies on user deanonymization. The shopping websites use cookies for multiple purposes such as storing information about the user including their shopping behavior and payment methods. The cookies may reveal the following information about the user to the third parties: the time of the payment, the payment address, the price, and personally identifiable information (PII), which refers to the information that reveals the real-world identity of the user (e.g., name, email, or phone). It is evident that the shopping website can easily deanonymize the user as it has access to all the information above. Thus, it is assumed that the attack is conducted by third-party trackers. Two attacks are studied: single transaction linkage, and cluster intersection. In the single transaction attack, the third-party attempts to deanonymize the user based on a single transaction generated by the user. Attack success is dependent on the amount of exposed data by the shopping website, where greater data exposure leads to higher attack success. The attacker can even identify the user by knowing the price of the item or the time when the user purchased an item, particularly when multiple transactions are examined.

The cluster intersection attack is complementary to the single transaction linkage attack. The tracker creates clusters of users based on their shopping behavior and attempts to gather more information about the user in order to deanonymize the user. Thus, this attack uses the output of the first attack as the input. To evaluate the success rate in identifying the users, 25 purchases were made on 20 merchant websites. To enhance anonymity, fresh keys produced by a mixing service were employed. A mixing service enables the users to change the address of their coin and thus break linkability. The results showed that, in 20 of 25 purchases, the third party was able to identify the user.

Reid and Harrigan studied user deanonymization using off-the-chain information [17]. To deanonymize the users, the malicious nodes first construct transaction and user network graphs as outlined above. By analyzing the graphs, the malicious node can identify multiple PK^+ that may potentially belong to the same user. The malicious node then utilizes off-the-chain information to deanonymize such users. Many organizations collect and store information about the real-world identity of the user (e.g., credit card information, address, name). Thus, the malicious node may attempt to access such information by eavesdropping on the communications between the user and such organizations. The underlying IP address of anonymous users may also be revealed using the information some organizations provide publicly. For instance, Bitcoin Faucet is a website that accepts donations from users and distributes those in small amounts between the recipient organizations. To assure the users that the money is donated, the company reveals the IP address of the nodes that received donations. The authors show that the IP can be linked to the transactions in the blockchain and thus can be used to deanonymize such users. The malicious node may also rely on the voluntary disclosure of some information (e.g., a user may reveal its identity in a public forum). Using these methods, the authors succeed in identifying malicious nodes that were involved in the theft of 25,000 BTC.

In this section, we studied multiple threats against the anonymity of the users in blockchain-based frameworks. The studies prove that the malicious nodes can employ multiple methods to deanonymize the users using diverse information sources that are stored on or off the chain or inferred from network activity. Next, we discuss methods for enhancing the anonymity level.

7.3 Protecting the User Anonymity

In this section, we outline some fundamental methods employed in the literature to protect the anonymity of the users in blockchain-based systems. As outlined in Section 7.2, the malicious nodes may link different transactions generated by the same user to deanonymize the user. To enhance the level of anonymity, in most existing blockchains, the transaction generator changes its PK^+ per transaction. The new address of the node is known as a shadow address. However, as outlined in Section 7.2.1, malicious nodes can classify new addresses and link them to the original address of the user and thus attempt to deanonymize the user.

The addresses employed by a user is maintained and managed by a wallet software that is installed on the user phone or computer. The wallet, among other things, stores the PK^+ and PK^- of the user. Some studies have suggested installing a new wallet software for different purposes or for communicating with different group of participants while interacting in the same blockchain [18] as a means to hide other identities of the user and enhance anonymity. The users also must not reveal information about their PK^+ in off-chain resources (e.g., forums) [8]. To further enhance the degree of anonymity of the users, other works have suggested employing more complex methods compared to those outlined above. We classify these methods in two categories: mixing services, and cryptographic methods, which are summarized in Table 7.2. Observe that the anonymity of the user comes with an extra cost in the form of computational resources or processing delay. In the rest of this section, we elaborate on these methods in detail.

7.3.1 Mixing Services

As outlined earlier in Section 7.2, the malicious entities try to link multiple transactions generated by a user to deanonymize the user. To enhance the unlinkability of the transactions, mixing services obfuscate the links between the transactions by mixing the input and output of transactions and generating completely new random transactions. The mixing service is intended to confuse the history of coin exchange. An ideal mixing service should offer the following key features [19]:

- *Accountability:* In the case of misbehavior by the mixing service provider, the users shall be able to prove the misbehavior and recover their funds.

Table 7.2
A Summary of the Methods to Enhance Anonymity of the Users

Method	Features	Limitations
Change key per transaction	The node uses a fresh key in each transaction	High degree of linkability
New wallet	New wallet is employed for groups of transactions	High degree of linkability, incurs overhead for managing wallets
Centralized mixing services	A central authority mixes the history and address of coins	Central controller may be able to track identities, suffers from centralization, medium degree of linkability
Decentralized mixing services	Nodes mix the history and address of coins in a distributed manner	Incurs delay, medium degree of linkability
Cryptographical methods	Mix the history of coins by encrypting input/output values	Incurs processing overhead
Timestamp obfuscation	Nodes generate the transactions with different timestamps as the original packets	Requires buffering the transactions

- *Anonymity:* The old address and the new address of the user must only be known to the user.
- *Resilience to denial of service (DOS) attack:* The mixing service shall remain secure against attacks such as DOS where the malicious entity attempts to make the service unavailable to users.
- *Scalability:* The mixing service must scale for large numbers of inputs and outputs.
- *Incentive:* The mixing service provider must receive incentive for the resources allocated to the task. The incentive must also remain fair to the end users.
- *Backward compatibility:* The mixing service must remain compatible with the existing blockchain protocols.

The mixing can be conducted by either a central mixer server or a group of decentralized nodes as outlined next.

7.3.1.1 Centralized Mixing Services

In this method, a centralized trusted node functions as the mixing server as shown in Figure 7.3. The participating nodes pay a service fee to the mixing server. The mixing server shuffles the inputs and outputs of the transactions and

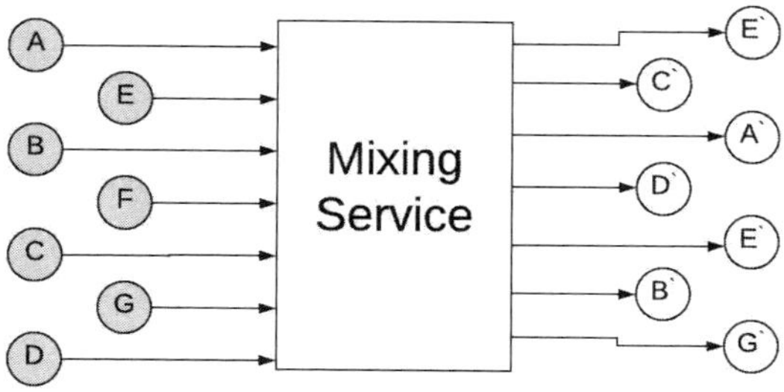

Figure 7.3 A high-level view of a central mixing server.

pays the received amount of coin back to the payers. The mixing services can be categorized as explicit and hidden based on the visibility of the transaction inputs and outputs to the central mixer. In the explicit models, the mixer knows the exact values employed as input and output and records the previous and new PK^+. This potentially introduces a privacy risk for the users as the mixer can track the user. However, the user will remain anonymous to other participants in the network. To address this challenge, in hidden methods, the values are hidden from the mixer.

The authors in [19] introduced a mixing service that relies on blind signatures to hide the input and output values. Blind signature is a cryptographic algorithm enabling users to sign a message, while the content of the message remains secret to the participants. The nodes can verify the signature without requiring access to the actual message, which introduces high privacy for the users [20]. The message can be unblinded at any time by the message generator. Initially, the mixing service (referred to as mix) publishes a set of mix parameters to a mixing log that is publicly available to all participants. The mix parameters include information about the service including the time instants when certain steps involved in the mixing are to be performed. A user selects the service and sends a message to the mix that contains *m* = {*mix-parameters, T*[*Key, n*]} where *T*[] is a token that contains the output public key and a nonce value used to determine the mixing fee. To protect anonymity, the token is encrypted using a commitment function that is known only to the user.

The user then sends *m* to the mix. The mix then sends a warranty back to the user that includes *c* = {*T, escr, mix-parameters*} and is signed with PK^- of the mix. *T* is the token received from the user and *escr* is the address to which the user must pay the coins. Upon receipt of *c*, the user pays the *escr* address. The mix then signs the token and sends a confirmation to the public log that the user has paid the mix. To claim the coin back, the user applies the commitment function on the token that reveals *T* = [*Key, n*]. Next, the user unblinds *Key* by sending the signed token to the public log file. The latter is used to verify the output by the mix as a proof of knowledge. The mix runs a beacon function to determine the mix fee and then transfers the funds to *Key*.

The mixing services are also available online through mixing service websites [21–23].

Similar to the above work, the authors in [24] introduced Mixcoin, which eliminates tokens. Thus, the warranty is as follows: $c = \{key, n, escr, mix\text{-}parameters\}$, which potentially reduces the complexity of the framework.

The authors in [25] introduced CoinJoin, a transaction formation method where multiple nodes jointly generate a single transaction. The nodes agree on the total input and output of the transaction. Each node independently signs the transaction and identifies the output addresses. This potentially shuffles the input addresses and thus enhances the anonymity of the users. To increase the level of anonymity, the input value shall be equal for all inputs. CoinJoin can be implemented in both centralized and decentralized setting. In the centralized setting, the participating nodes request a central controller that acts as the trusted authority to transfer funds. The central authority constructs a new transaction and includes a number of participating nodes as the input. The authority sets the address of the nodes as the output. We will study the decentralized setting in the next section.

In centralized mixing services, the central authority remains a bottleneck that limits scalability. Additionally, the central authority may attempt to deanonymize the users or record the new and old PK^+ associated with a user. To address these limitations, decentralized mixing services have been introduced.

7.3.1.2 Decentralized Mixing

In decentralized mixing model, the participating nodes jointly conduct the mixing task without relying on a central trusted third party (TTP). In the decentralized setting of CoinJoin, the participating nodes agree to generate a joint transaction. The transaction is then generated by all the participants where each one populates the signature corresponding to its own input and provides its own output. In this method, the nodes participating in the transaction may be able to track the old and new inputs and outputs of the transactions. However, this method requires the nodes to trust each other. To address this challenge, blind signature is used as it inherently guarantees the honest behavior of the participants (see Section 7.3.1.1).

CoinShuffle is a decentralized mixing service where the participating nodes initially join a public bulletin board that facilitates identifying other nodes that wish to mix their transactions [30]. A group of nodes then jointly start the shuffling process. Each node creates a new output address and adds it to the transaction. To prevent another node from tracking the node output address, the node shuffles the outputs of the transaction. Then the transaction is sent to another node in the group. The process is repeated until all nodes add their output address to the transaction. Each node checks if its new output is included in the transaction. If so, each node generates a mixing transaction that spends coins from all input coins to the shuffled addresses.

Xim [26] is a decentralized mixing service that employs a fair exchange protocol to enhance the anonymity of blockchain users. Assume that Alice wishes to mix coins in a distributed manner. She first expresses her willingness by broadcasting a transaction in the blockchain. The nodes that are interested in participating directly communicate with her. Alice then selects one node randomly and sends a confirmation to this node (say, Bob). The selected participant is not visible to the blockchain network, which increases the user anonymity and prevents other nodes from linking transactions. Alice and Bob first exchange their new address and then transfer coins to the new address of the other party using the fair exchange protocol. The fair exchange protocol ensures that the participants commit to their obligations without relying on a centralized trusted party. At the end of the mixing process, Alice can broadcast another mixing request. Alice may repeat these steps as many times as she wishes to enhance her anonymity level. Note that generation of transactions involves a transaction fee; thus, there is a trade-off between the anonymity level and cost. Xim leaves no traces of the mixing process on the blockchain, which makes it difficult for the attackers to link the transactions of the users.

CoinSwap is another decentralized mixing service where mixing is executed between three participants [27]. Figure 7.4 summarizes the CoinSwap process. Assume that Alice, Bob, and Carol aim to mix their keys. Initially, Alice computes a 2-of-2 escrow with Bob that refers to an escrow that requires 2 transactions to pay the screw and 2 transactions to release it. This is transaction is then sent to Bob. Similarly, Bob computes a 2-of-2 escrow with Carol. Once these transactions are broadcast in the network, Bob can be assured that if he pays Carol, he will receive the coin from Alice. With this assurance, Bob can proceed to transfer the coin to Carol. Alice also pays Bob. The outputs are new identities that potentially enhance the anonymity of the users.

Although mixing services are useful to break the link between multiple transactions of a user, they are vulnerable to malicious mixers who might steal the user coin or track the transactions of the users. This limits the level of anonymity offered by such services. Cryptographic methods that address these limitations are discussed next.

7.3.2 Cryptographical Methods

In these methods, the anonymity of the user is guaranteed cryptographically. Blind signature is an example of a cryptographic method where the users sign transactions without knowing the exact content of the transaction. Blind sig-

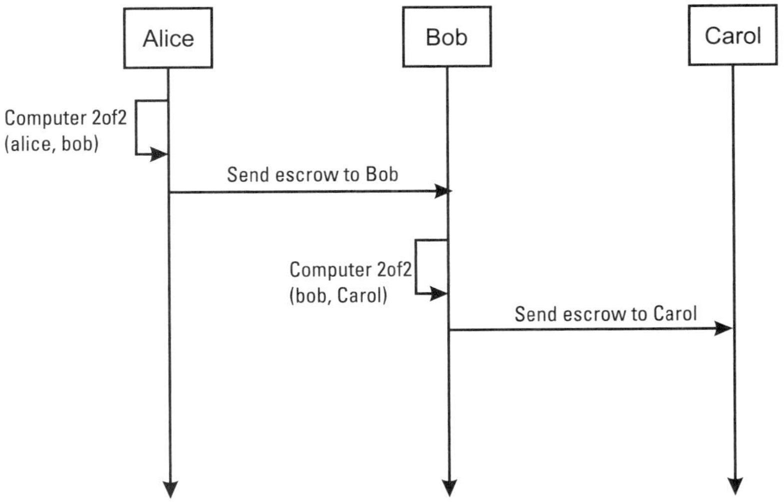

Figure 7.4 A high-level view of the process in CoinSwap.

nature is widely used with mixing services to enhance anonymity as outlined in the previous section.

ZeroCoin is among the first methods to improve the anonymity of the users in Bitcoin [28]. The user selects a random serial number and encrypts the serial number using a commitment algorithm. To mine ZeroCoin, the user sends the output of the commitment algorithm along with an equal amount of Bitcoin money to the network. To spend the transaction, the user must prove that he or she previously generated a ZeroCoin and knows the corresponding serial number. This is achieved using zero knowledge proof. Using the latter, a node can prove knowledge of a particular data without revealing the data to the verifiers. ZeroCoin is compatible with Bitcoin and thus can be implemented on top of Bitcoin.

The authors in [29] introduced a new algorithm based on composite signature where multiple transactions are combined into one transaction to obfuscate the links between the input and outputs. Based on the composite signature, a number of individual signatures $PK^{-}_{1}(m_1)$, $PK^{-}_{2}(m_2)$, $PK^{-}_{3}(m_3)$, ..., $PK^{-}_{i}(m_i)$ are combined into a composite signature $C\text{-}PK^{-}_{\{1,2,3,..,i\}}(m_1, m_2, m_3, \ldots, m_i)$. Once a composite signature is generated, new signatures can be added at any time. To enhance anonymity, each user leaves a particular amount of coin as joining fee in a transaction that is not payed to any node. A merging service is run in the network that collects all transactions in the network. Once the number of collected transactions reaches a predefined value, the merging service combines all of them in a single consolidate transaction using composite signature, which reduces the public record available about each user. The transaction is then sent to one peer in the network. The receiving peer adds its PK^+ as the output of the transaction by claiming the joining fee, which enhances the user anonymity.

Another method to shuffle outputs and ensure anonymity is the Ring signature, which is a group signature method requiring that each participant maintains some of the keys and that all participating nodes sign the message. CryptoNote [30] employs the Ring signature to enhance the user anonymity. Assume that Alice is going to pay Bob. Alice generates a one-time PK^+ using Bob's PK^+ using existing cryptographic methods and uses the one-time key as the output of the transaction. Bob verifies the transaction using the PK^- corresponding to the one-time key. The transaction may also include additional input and output. The one-time key is unlinkable to Bob's previous transactions, as it is impossible to find which input and output belong to Bob.

In this section, we studied the key methods employed in the literature to enhance the user anonymity. As multiple studies suggest, to enhance the anonymity, the users may employ a combination of methods (e.g., a mixing service that uses blind signature to hide the input and output values). In the next section, we study key management methods that are used to manage the keys used as the identity of the users in the blockchain.

7.4 Key Management

In cryptocurrencies, the users can change their key in each transaction such that their remaining funds are transferred to the new PK^+. The users can thus remove the old key pair and only maintain the new key pair. Users may decide to distribute their coins among multiple transactions to enhance their anonymity. In this instance, the users would be required to store multiple key pairs. It is rather important that these key pairs are stored in a secure manner, emphasizing the need for an effective key management solution.

Unlike cryptocurrency, where the node does not require storing the old keys, in CPS, the devices or the users may need to maintain the old key pairs to be able to prove ownership of a particular transaction at a later point in time. Considering the large volume of transactions and the number of devices owned by a user, there may be a need to manage a significant number of key pairs, which makes key management in blockchain-based CPS even more challenging than cryptocurrencies. The existing key management solutions in blockchain are categorized in six groups as follows [31]:

1. *Local storage:* The blockchain users can store keys in a local database and configure the path to the file where the keys are stored in the wallet software. The wallet software automatically reads keys and gener-

ates and stores new keys in this local storage. The main advantage of this method is that it offloads the key management task to the user device. The users can generate as many keys as they wish. However, this method also introduces security and anonymity concerns. If the user device is compromised, then all addresses of the user are revealed. Malware installed on the device may also attempt to read the keys to deanonymize the users.

2. *Encrypted wallets:* Some wallets allow the user to encrypt the local storage file using a password chosen by the user. This potentially enhances the security of the keys against physical theft; however, it does not provide security against digital attacks (e.g., malware can still read the file using the access permissions of the user).
3. *Offline key storage:* In this method, the keys are stored in an offline storage (e.g., an external hard disk drive (HDD)). However, it impacts the accessibility of the keys as the wallet software does not always have access to the keys. Printing key pairs in papers or converting them to Quick Response (QR) barcodes can provide ease of access for users. However, there are some serious security implications associated with these conveniences.
4. *Air-gapped key storage:* Similar to offline key storage, in this method keys are stored in a separate offline storage that has access to the keys and can generate or read keys and generate new transactions. The transactions are then transferred to another device that is connected to the internet to be transferred to the blockchain. This method adds an additional security layer that further protects the key security.
5. *Password-derived keys:* In this method, instead of storing keys, the keys are constructed using a password that is only known to the user. However, for each key pair, the user must create a new password. To address this challenge, a hierarchical deterministic (HD) wallet [32] is introduced, which enable the users to generate multiple child keys by knowing a master key. In the HD wallet, the sequence of child private keys (denoted as d_i) are constructed using a master private key (d`) using the following:

$$d_i = hash(i, d`)$$

Although the HD wallet enables the users to construct multiple keys using a single parent key, studies have shown that the HD wallet is vulnerable to malicious nodes that attempt to derive the parent private key using the parent public key [32]. The authors in [32] addressed this challenge by proposing an optimized version of the HD wallet that employs m master private keys in place of a single key. This method tolerates the leakage of up to m keys meaning that if up to m keys are compromised, the malicious node still cannot construct all the child keys, which enhances the security of the HD wallet.

In [33], we introduced the concept of Generator Verifier (GV) to address the key management challenge. GV enables the participating nodes to manage all their transactions using a single key pair and a secret value. A detailed discussion on GV was covered in Section 5.3.2.1.

6. *Hosted wallets:* In this method, the keys are stored in a third-party web service that enables the users to store all their keys and retrieve them on-demand. The web service employs conventional authentication methods that include passwords or two-step verification. However, these third-party web services will remain an attractive target to the malicious nodes. The malicious nodes may also attempt to eavesdrop on the communications between the web service and the users to gather information necessary to deanonymize the users (e.g., the IP address of the users).

In summary, this section outlines the existing solutions to address key management in the blockchain. Managing keys is critical for enabling nodes to change their identity and thus achieve a higher level of anonymity. Next, we study the anonymity of the nodes in CPS applications.

7.5 Anonymity in CPS

In this section, we discuss the anonymity of the blockchain users in the context of CPS. The transactions in CPS applications may include data or contain a link to data stored in a cloud storage. The input and output of such transactions also vary depending on the application (e.g., a smart contract may be generated to facilitate trade of goods between the users). Any future transaction related to the trade uses the address of the smart contract as the input. In this section, we study the anonymity of the CPS users in a blockchain. The main aim of the attacker in this study is to classify the type of devices installed in a site. In a conventional CPS setting, device classification is a security measurement that enables the network administrators to identify the devices installed in the site and thus detect unauthorized devices [34–36]. However, in CPS, device classification introduces privacy concern as the malicious node can identify the type of devices installed in the user site. Attackers can then infer user activity and potentially uncover the user's identity.

In the rest of this section, we first provide a high-level overview of our framework to study user anonymity in CPS in Section 7.5.1. Then we discuss the attack model in Section 7.5.2. We outline the methods to protect the anonymity of the user in Section 7.5.3. Finally, the experimental results are discussed in Section 7.5.4.

7.5.1 Overview

To study the anonymity of the users in the blockchain, we first populate a blockchain based on transactions corresponding to the communications between the devices in a smart home. We use a smart home setting as a representative case; however, the concept can be extended to any other application. We used the smart home traffic data set available in [37] that consists of network traffic data collected over 2 weeks for 28 off-the-shelf IoT devices, for example, cameras, lights, plugs, motion sensors, and appliances (referred to as the test set in the rest of the chapter). The test set includes the data of any communication the devices made with the user, other devices, or the service providers along with the network related traffic (e.g., simple mail transfer protocol (SMTP) packets). The communications between the devices are represented using transactions in the blockchain. The structure of a transaction is as follows:

$$T_{ID} \| P.T_{ID} \| TimeStamp \| Output \| PK^{+} \| Signature$$

where T_{ID} is the identity of the current transaction, $P.T_{ID}$ is the identity of the previous transaction generated by the same node, and *TimeStamp* represents the time when the device initiated the communication. We use the timestamp in the packets in the dataset to populate this field. *Output* is the hash of the PK^+ that will be used to generate the next transaction. This ensures that the PK^+ of each transaction differs with the other transactions and thus introduces a level of anonymity. PK^+ and *Signature* are populated by the transaction generator. In case the communication between devices involves data, the hash of the data is signed and stored in the *Signature* field. The transactions are collected and stored in the blockchain by a single miner. Given that the consensus algorithm does not impact the anonymity of the blockchain and our study, we disregard that process.

7.5.2 Attack Model

We use machine learning to model attacks that attempt to classify IoT devices. As shown in Figure 7.5, the pattern of transactions mostly represents a sequence of in-order numbers. Different patterns share some features (e.g., a

Device	Patterns of frequent time separation (in s)
Smart_Things	0.207 then 58 then 0.207 then 58
Amazon_Echo	0.217 then 30 then 0.004 then 30
TPLink_Camera	0.12 then 61 then 0.12 then 61
Samsung_Camera	0.165 then 30 then 0.165 then 30
Drop_Camera	1.03 or 0.2
Insteon_Camera2	9x<0.0001 then 0.216 then 300 ...
Baby_Monitor	600 then 0.28 then 600 then 0.28
TPLink_Smartplug	0.24 then 236 then 0.24 then 236
TPLink_Smartplug	0.12 then 236 then 0.12 then 236
iHome	60 then 0.205 then 60 then 0.205
Nest_Smockalarm	0.207 then 0.015 then 0.207 then 0.015
Netatmo_Weather	1.72 then 0.33 then 1.72 the 0.33
Sleep_Sensor	10 then 0.276 then 10 then 0.276
Lifx_Smartbulb	1.92 or 60
Triby_Speaker	120 0.3 - 120 0.3 - 56 0.3
Pix_Photoframe	0.31 or >=0.3 then 65 then 650
HP_Printer	90

Figure 7.5 Inter-packet temporal patterns for devices.

separation of 0.207s is found for both the SmartThings and the Nest smoke alarm). This pattern can be best represented by trees; thus, the machine learning algorithms, employed by the attacker, use decision trees to analyze the pattern of transactions in the blockchain and classify devices.

The attacker can read the blockchain information but cannot decrypt data without the associated PK^{-}. Initially, the attacker trains the decision tree algorithm on a local network, referred to as the testnet in the rest of this chapter. Note that the blockchain contains only the smart home communications; thus, the attacker can install some devices normally installed in the smart homes to build the testnet. The testnet identifies the ability of the attacker to classify devices. We study two attack models:

- *Informed attack:* In this attack, the attacker knows the exact number and type of devices that are installed in a smart home and thus can populate the testnet with transactions corresponding to the same range of devices, which lead to an accurate testnet. We model this attack using tenfold cross-validation analysis where the training process always ensures that all smart home devices are represented in the training algorithm. Although it is unlikely that, in a real scenario, the attacker would know all devices installed in a smart home, we consider this attack model as the worst-case scenario for device classification.

- *Blind attack:* In this attack, the attacker does not know the number or type of devices installed in the smart home; thus, the training set may contain more or fewer devices than the test set. Consider a realistic scenario: the attacker does not exactly know what devices are installed in the smart home, thus, the attacker can only guess what devices the users may have installed in their smart home and thus populate the testnet using those devices. The attacker creates his or her own lab that contains all, some, or even none of the devices in the testnet.

7.5.3 Protecting User Anonymity

In CPS-based blockchain, each device generates transactions in particular time intervals. Studies have shown that the malicious nodes may identify the devices using the pattern of the transactions even if the transaction content (i.e., the data of the devices) is encrypted [38]. To enhance the anonymity of the users in CPS-based blockchain, we propose multiple timestamp obfuscation methods that potentially enhance the resilience to device classification. To demonstrate the effectiveness of these methods, we compare them with a baseline, where each communication is represented as a transaction. The timestamp of the transaction equals the time when the communication packet is generated. Each device chains all its transactions to a single ledger and changes its PK per transaction. We propose three timestamp obfuscation methods:

1. *Delay transaction generation:* In this obfuscation method, the transaction corresponding to the communication of a device is generated after a random delay. Assuming that T_c represents the time when the communication took place, T_d represents the time when the transaction was generated; thus, $T_d = [T_c, T_{c+max\text{-}delay}]$, where *max-delay* is the maximum delay that can be applied in a transaction and is defined by the network manager. The random delay is independently defined for each transaction.
2. *Combine packets into a single transaction:* In this obfuscation method, multiple packets sequentially generated by a device are combined into a single transaction. In other words, a single transaction contains the hash of the exchanged information in a group of packets. This shares similarities with summarizing transactions in the blockchain as studied in Chapter 5. This obfuscation method enhances the anonymity of the user as the volume of data available in the public blockchain decreases and the transaction pattern stored in the blockchain differs from the real communication pattern.

3. *Combine transactions from multiple devices into a single ledger:* In this obfuscation method, we employ a single ledger to store transactions corresponding to multiple devices. In conventional blockchains, the transactions of individual devices are chained in the same ledger, which enables the attacker to get access to the full history of the interactions of a device. By sharing a common ledger across multiple devices, the pattern of transactions will change, which potentially enhances the resilience of the blockchain against device classification.

7.5.4 Experimental Results

In this section, we outline the results of the experiments to classify the smart home devices. We first evaluate the impact of the delayed transaction obfuscation method on the success rate in classifying the smart home devices. The results of the experimental study are shown in Figure 7.6. We applied random delay in [0, 0.5], [0, 2], and [0, 30] seconds, respectively. The results demonstrate at least 15% improvement of the classification accuracy in the informed attack, while in blind attack it shows a maximum of 10% improvement.

We next study the impact of grouping transactions from multiple nodes in one ledger. The experimental results are shown in Figure 7.7. We vary the number of devices per ledger from 2 to 10 devices. As evident from the results, in the informed attack the success rate of device classification reduces from about 98% to 63%. When a small number of devices share a ledger, it is easier to distinguish and thus classify devices. As more devices are combined in a ledger, this distinction is harder to make. Interestingly, in the blind attack, the success rate of device classification increases slightly as more devices store transactions in the ledger. With more devices storing transactions in a ledger, the number of devices similar in both training and the test set increases; thus, the success rate in classifying those increases.

Finally, we study the impact of multiple packets per transactions. The results are shown in Figure 7.8. As evident, in the informed attack, the obfuscation method reduces the success rate of classification by roughly 15% to 20%. The improvement is lower in the blind attack, as in this attack the attacker is only aware of a subset of the actual devices.

In summary, in this section, we demonstrated that by analyzing transactions in the blockchain, malicious nodes can classify CPS devices in a smart home setting. Successful device classification allows an attacker to use temporal patterns of device-specific transactions to infer the start, end, duration, and intensity of user activities, which, when linked to external data sources, can lead to deanonymization of the user.

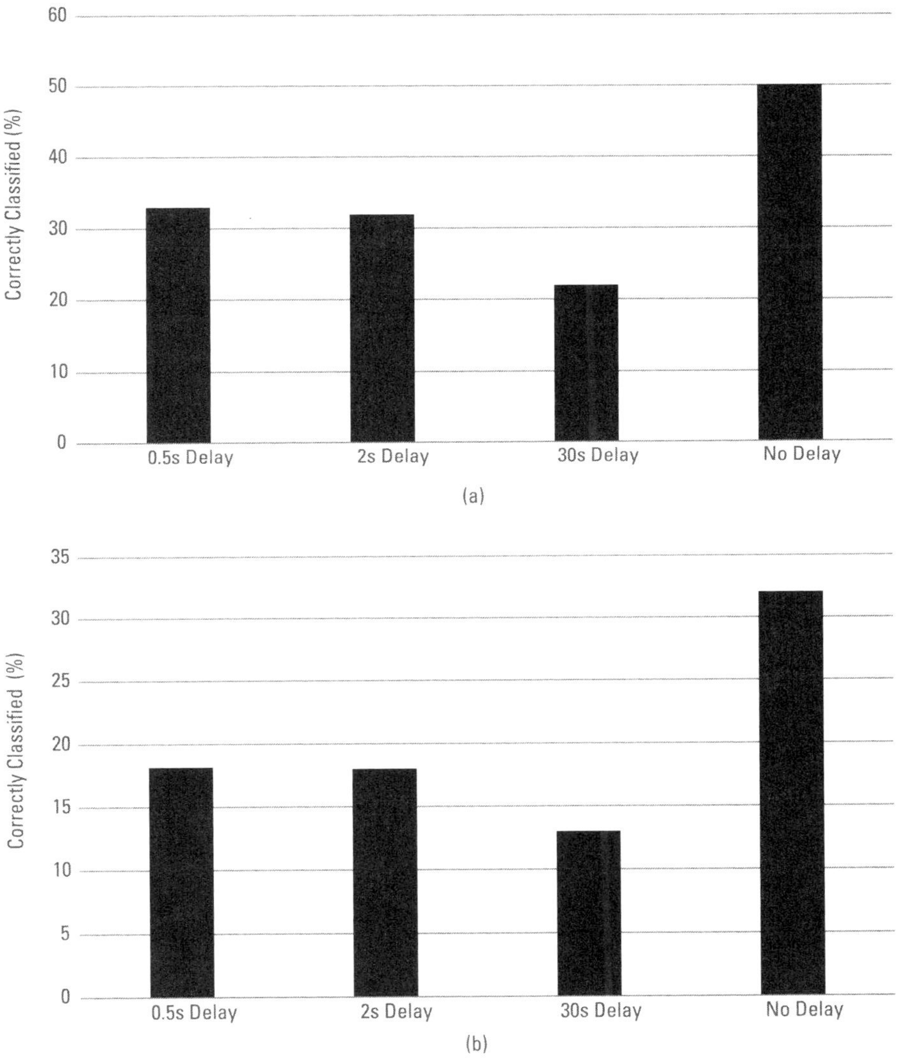

Figure 7.6 The impact of delayed transaction obfuscation for (a) informed attack and (b) blind attack.

7.6 Conclusions

Blockchain users employ a PK^+ as their identity that introduces a degree of anonymity, as the real-world identity of the user is not revealed to the blockchain participants. Malicious nodes may attempt to deanonymize the users by active interactions with target nodes, analyzing blockchain transactions, analyzing network traffic, and analyzing off-the-chain data. To enhance the anonymity

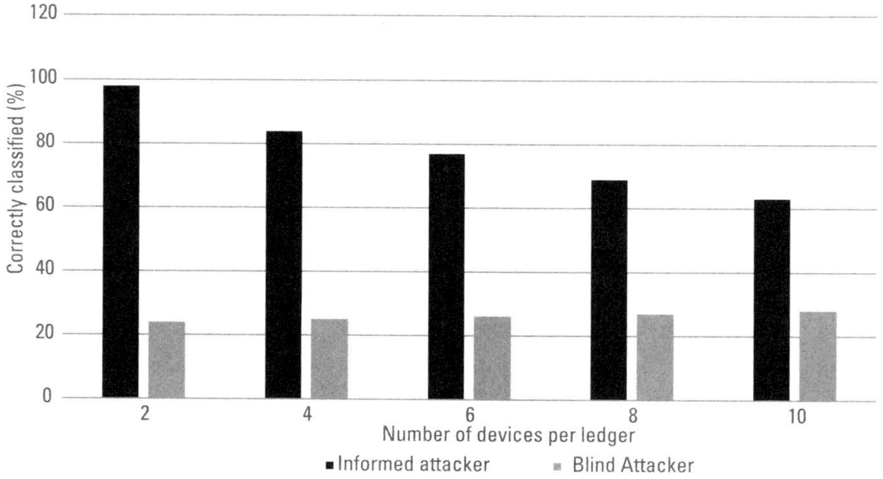

Figure 7.7 The impact of multinode per ledger in informed and blind attacks.

level, the blockchain users can change their PK^+ per each transaction, employ mixing services to break the link between the input and output, employ cryptographic algorithms to hide input and output values, and use multiple wallets. As the blockchain user may potentially use multiple keys, key management is a significant challenge. We studied the existing solutions to address this challenge. Finally, we studied the anonymity of the users in the CPS. We populated a blockchain using the communications of smart home devices and analyzed the resulting blockchain transactions using machine learning algorithms. The results demonstrated a high success rate in classifying the smart home devices. The classification is based on the pattern in which the transactions are generated; thus, we introduced three obfuscation methods to change the transaction patterns, which reduced the chance of the successful classification of devices.

Overall, supporting CPS anonymity with the blockchain is a multilayered challenge that needs to consider on-chain and off-chain data, network interactions, and temporal patterns. Vulnerabilities at any of these layers may compromise anonymity. Therefore, strong anonymity guarantees require well-integrated and robust approaches that protect against the vulnerabilities at all layers.

This chapter concludes Part II of the book, where we have explored specific yet application-agnostic challenges of applying the blockchain in CPS, namely, scalability, immutability, trust, and anonymity, as well as solutions to these challenges. In other words, Part II has taken a horizontal perspective on the blockchain for CPS by exploring these cross-cutting challenges and solutions. Part III of the book, which includes Chapters 8 to 11, takes a vertical

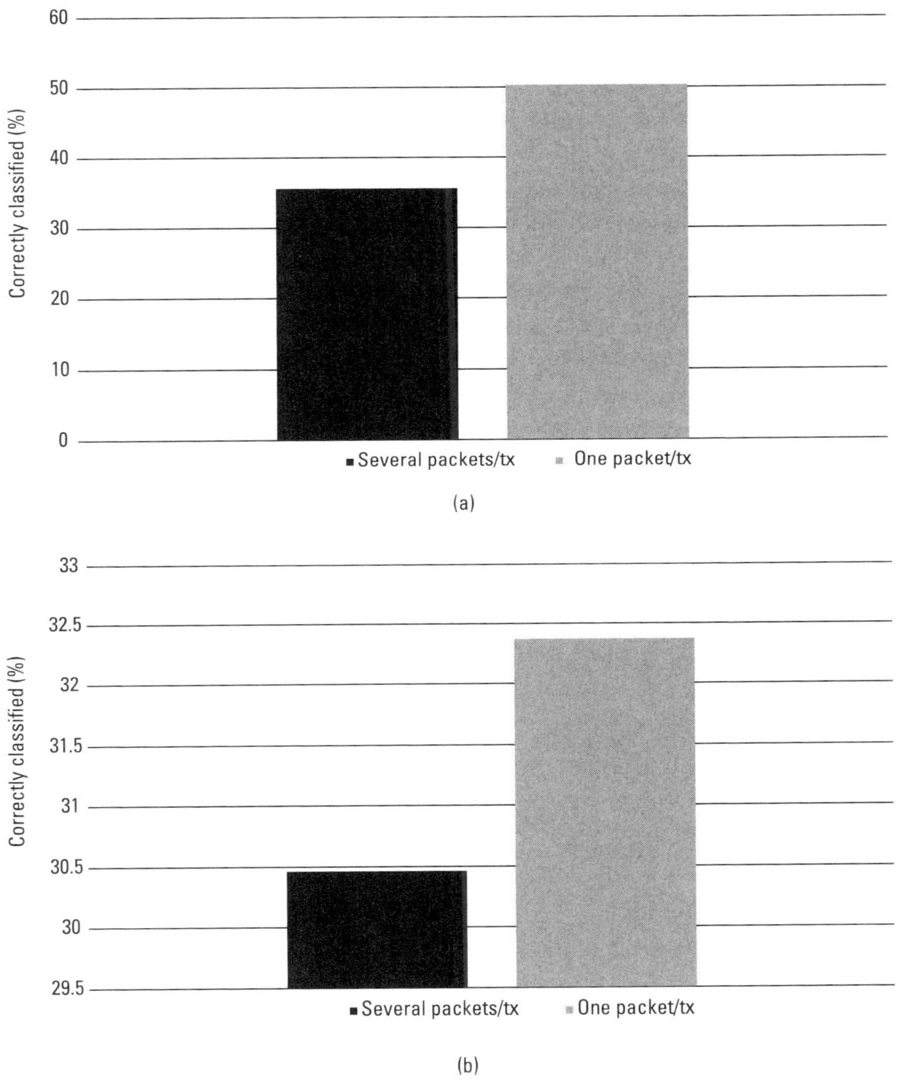

Figure 7.8 The impact of multipacket transactions for (a) informed attack and (b) blind attacks.

perspective on the blockchain for CPS. Each chapter in Part III focuses on a specific application area, ranging from smart grids (Chapter 8), smart vehicles (Chapter 9), supply chains (Chapter 10), and IoT data marketplaces (Chapter 11). We will see how each application area presents some additional unique challenges that require tailored blockchain solutions for CPS.

References

[1] Bradbury, D., "Anonymity and Privacy: A Guide for the Perplexed," *Network Security*, Vol. 10, 2014, pp. 10–14.

[2] Nakamoto, S., *Bitcoin: A Peer-to-Peer Electronic Cash System*, Manubot, 2019.

[3] Amarasinghe, N., X. Boyen, and M. McKague, "A Survey of Anonymity of Cryptocurrencies," *Proceedings of the Australasian Computer Science Week Multiconference*, January 2019, p. 2.

[4] Feng, Q., et al., "A Survey on Privacy Protection in Blockchain System," *Journal of Network and Computer Applications*, Vol. 126, 2019, pp. 45–58.

[5] "Mt. Gox," https://btc.com/stats/pool/Eligius, accessed April 2020.

[6] Meiklejohn, S., et al., "A Fistful of Bitcoins: Characterizing Payments Among Men with No Names," *Proceedings of the 2013 ACM Conference on Internet Measurement Conference*, October 2013, pp. 127–140.

[7] "Tor Project," https://www.torproject.org/, accessed April 2020.

[8] Khalilov, M. C. K., and A. Levi, "A Survey on Anonymity and Privacy in Bitcoin-Like Digital Cash Systems," *IEEE Communications Surveys & Tutorials*, Vol. 20, No. 3, 2018, pp. 2543–2585.

[9] Biryukov, A., and I. Pustogarov, "Bitcoin over Tor Isn't a Good Idea," *2015 IEEE Symposium on Security and Privacy*, May 2015, pp. 122–134.

[10] Koshy, P., D. Koshy, and P. McDaniel, "An Analysis of Anonymity in Bitcoin Using P2P Network Traffic," *International Conference on Financial Cryptography and Data Security*, March 2014, pp. 469–485.

[11] Fanti, G., and P. Viswanath, "Anonymity Properties of the Bitcoin P2P Network," *arXiv preprint arXiv:1703.08761*, 2017.

[12] Biryukov, A., D. Khovratovich, and I. Pustogarov, "Deanonymisation of Clients in Bitcoin P2P Network," *Proceedings of the 2014 ACM SIGSAC Conference on Computer and Communications Security*, November 2014, pp. 15–29.

[13] Conti, M., et al., "A Survey on Security and Privacy Issues of Bitcoin," *IEEE Communications Surveys & Tutorials*, Vol. 20, No. 4, 2018, pp. 3416–3452.

[14] Androulaki, E., et al., "Evaluating User Privacy in Bitcoin," *International Conference on Financial Cryptography and Data Security*, April 2013, pp. 34–51.

[15] DuPont, J., and A. C. Squicciarini, "Toward De-Anonymizing Bitcoin by Mapping Users Location," *Proceedings of the 5th ACM Conference on Data and Application Security and Privacy*, March 2015, pp. 139–141.

[16] Goldfeder, S., et al., "When the Cookie Meets the Blockchain: Privacy Risks of Web Payments Via Cryptocurrencies," *Proceedings on Privacy Enhancing Technologies*, No. 4, 2018, pp. 179–199.

[17] Reid, F., and M. Harrigan, "An Analysis of Anonymity in the Bitcoin System," in *Security and Privacy in Social Networks*, New York: Springer, 2013, pp. 197–223.

[18] "Bitcoin User Privacy," https://bitcoin.org/en/protect-your-privacy.

[19] Valenta, L., and B. Rowan, "Blindcoin: Blinded, Accountable Mixes for Bitcoin," *International Conference on Financial Cryptography and Data Security*, January 2015, pp. 112–126.

[20] Chaum, D., "Blind Signature System," *Advances in Cryptology*, Boston, MA: Springer, 1984, p. 153.

[21] "Bitcoin Fog," https://bitcoinfog.info.

[22] https://en.bitcoinwiki.org/wiki/Bitmixer.io.

[23] "Bitmixer," https://helixmixer.org/helix/light.html.

[24] Bonneau, J., et al., "Mixcoin: Anonymity for Bitcoin with Accountable Mixes," *International Conference on Financial Cryptography and Data Security*, March 2014, pp. 486–504.

[25] "CoinJoin," https://bitcointalk.org/index.php?topic=279249.0.

[26] Bissias, G., et al., "Sybil-Resistant Mixing for Bitcoin," *Proceedings of the 13th ACM Workshop on Privacy in the Electronic Society*, November 2014, pp. 149–158.

[27] "CoinSwap," https://bitcointalk.org/index.php?topic=321228.

[28] Miers, I., et al., "Zerocoin: Anonymous Distributed E-Cash from Bitcoin," *2013 IEEE Symposium on Security and Privacy*, May 2013, pp. 397–411.

[29] Saxena, A., J. Misra, and A. Dhar, "Increasing Anonymity in Bitcoin," *International Conference on Financial Cryptography and Data Security*, March 2014, pp. 122–139.

[30] "Cryptonote," https://cryptonote.org/whitepaper.pdf.

[31] Eskandari, S., et al., "A First Look at the Usability of Bitcoin Key Management," *arXiv preprint arXiv:1802.04351*, 2018.

[32] Gutoski, G., and D. Stebila, "Hierarchical deterministic Bitcoin wallets that tolerate key leakage," *International Conference on Financial Cryptography and Data Security*, January 2015, pp. 497–504.

[33] Dorri, A., S. S. Kanhere, and R. Jurdak, "MOF-BC: A Memory Optimized and Flexible Blockchain for Large Scale Networks," *Future Generation Computer Systems*, Vol. 92, 2019, pp. 357–373.

[34] Feng, X., et al., "Acquisitional Rule-Based Engine for Discovering Internet-of-Things Devices," *27th {USENIX} Security Symposium ({USENIX} Security 18)*, pp. 327-341.

[35] Miettinen, M., et al., "IoT Sentinel: Automated Device-Type Identification for Security Enforcement in IoT," *2017 IEEE 37th International Conference on ICDCS*, 2017, pp. 2177–2184.

[36] Sivanathan, A., et al., "Classifying IoT Devices in Smart Environments Using Network Traffic Characteristics," *IEEE Transactions on Mobile Computing*, 2018.

[37] "IoT Dataset," https://iotanalytics.unsw.edu.au.

[38] Apthorpe, N., D. Reisman, and N. Feamster, "A Smart Home Is No Castle: Privacy Vulnerabilities of Encrypted IoT Traffic," *arXiv preprint arXiv:1705.06805*, 2017.

Part III

8

Blockchain Applications in Smart Grids

Fengji Luo and Gianluca Ranzi,
School of Civil Engineering, The University of Sydney, Australia

8.1 Introduction

A significant emerging application for blockchain in CPS involves smart grids, where the increasing penetration of smart meters and distributed renewable sources require new decentralized architectures for energy management. From an architectural perspective, smart grids are CPS that include: (1) energy sources and consumers, which generate and consume energy; (2) smart meters, which are the sensing devices that observe the physical processes of energy generation or consumption; (3) the power grid, which is the cyberphysical network through which energy is transported; and (4) energy managers, which are resourceful devices that coordinate the actions of energy producers and consumers across a spatial region within the smart grid. Smart grids are a relatively unique class of CPS, where the objective is ultimately to track the flow of electrons, which are a pooled physical commodity that are exchanged across the grid in a large volume to provide electricity services. The pooling of electrons and their virtual similarity presents interesting challenges for trust, anonymity, and scalability, as it makes it more challenging for participants to ascertain the occurrence of events such as the transmission or reception of a set amount of energy. This raises the possibility, for instance, of malicious smart grid participants double counting energy that they sell to maximize their profits. This is in contrast to the trade of discrete physical commodities supported by blockchain and CPS (discussed in Chapter 10) or discrete digital commodities (discussed in Chapter 11), which involve their own distinct challenges. Before delving into how blockchain is used in smart grids, we provide a more detailed background on smart grids next.

Human society's energy demand has been experiencing a rapid increase in recent years. According to the "BP Statistical Review of World Energy 2019" [1], the global energy consumption demand grew at a rate of 2.9% in 2018, almost doubling its 10-year average of 1.5% per year, and the fastest since 2010. The ever-increasing energy demand, together with the associated climate change crisis, has been attracting tremendous attention from both academia and practitioners with a view towards developing solutions that can reshape human society's energy generation and consumption patterns and, therefore, supporting sustainable development.

Power grids are a fundamental infrastructure that transmit and deliver electricity, traditionally generated from fossil-fueled power plants, to end energy consumers. In the early twenty-first century, the notion of smart grids [2] was proposed to facilitate energy distribution and management. A broad range of CPS devices, including sensors, smart meters, and smart appliances, is incorporated in the smart grid and provides information about energy flow and grid operation and maintenance. This information is utilized by the grid operator to prevent any service disruption and ensure that the transmission lines are never overloaded. The two key characteristics that distinguish smart grids from traditional power systems are:

1. High penetration level of distributed renewable energy: These are the most common distributed energy sources (DESs) and include wind turbines and photovoltaic (PV) solar panels. The DESs are geographically disperse and are usually installed at the energy consumer's side, such as building rooftops and roads in urban areas. The highly penetrated DESs are significantly influencing traditional power systems to consider a move from their centralized arrangement to a distributed electricity generation approach. In this process, many traditional pure power consumers now have the capability of generating electricity and of injecting it into the grid.
2. Advanced metering infrastructure and active participation of end energy users: In traditional power systems, the information flow is one-way, and the electricity generation is completely driven by electricity consumption. The utility collects the consumer-side electrical load demand and schedules the power generators to produce the required amount of power to serve the load. In smart grids, such one-way information flow is transformed into a two-way information exchange,

supported by the deployment of smart meters. Smart meters are embedded within the communication modules and they automatically send the power consumption of the end users to the utility while providing incentives and pricing signals to end users (e.g., real-time electricity pricing data). The aim of the incentives and pricing signals encourage end users to reshape their electricity consumption patterns and assist the utility to achieve certain grid-level objectives, such as the reduction of the load during the peak hours. In this way, the end user in the smart grid is not just passively served by the grid but is actively participating in the operation of the grid, which is known as the demand response (DR) in the literature [3].

The transition from traditional power systems to smart grids increases the operational complexity of the grid and provides opportunities to create new interactions among end users. The prevalence of distributed renewable energy sources and sensing facilities has transformed the smart grid to a highly distributed environment hosting numerous geographically dispersed energy entities. This transition raises the need for an information infrastructure with the following features to support the operation of the modern grids:

- Secure and trustworthy: The ever-complex structure of modern grids increases its vulnerability to cyberphysical attackers. A robust, trustworthy, and attack-resistant information infrastructure is therefore needed to enhance the secure and reliable operation of the grids.
- Privacy aware: In a distributed environment, it is possible for various entities to collect privacy-sensitive information about the users (e.g., the energy consumption or generation patterns). Thus, it is critical to consider the privacy of the users while sharing data.
- Scalable: A scalable information infrastructure is desired that can serve the integration and interaction of a large number of energy resources that could be geographically distributed with diverse capacities and physical parameters.

Blockchain, being a distributed database with decentralized, trust, and traceability features (see Chapter 3), is a promising cyber infrastructure solution that would fit the distributed energy environment of the smart grid, as shown in Figure 8.1. Dong et al. [4] presented a comprehensive overview on the vision of integrating blockchain and other information infrastructure (i.e., IoT and cloud computing systems) into the future energy systems. The authors suggest that two fundamental services can benefit from this integration. Data

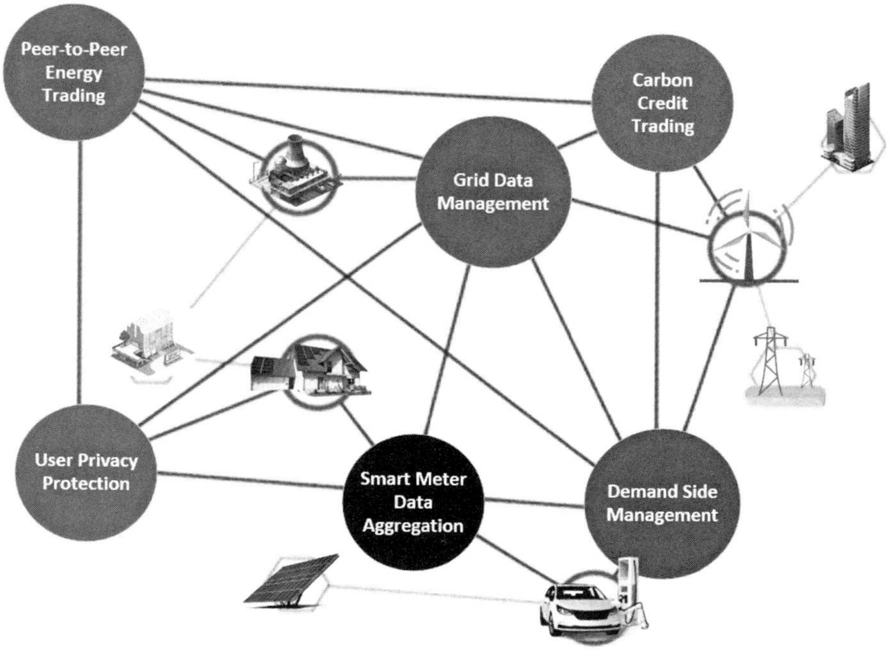

Figure 8.1 Vision of blockchain in the smart grid.

management refers to using blockchain to store data produced by the smart grid participants, which ensures the integrity of the data. The smart grid participants can trust that the historical data received from other parties remained unchanged since generation as the hash is stored in the blockchain, and authentication that refers to the use of blockchain to authenticate the participants in the smart grid. In addition to the outlined services, the emerging applications in the future grid can also benefit from the integration that include: (1) open energy market architecture, the use of blockchain to support energy trading activities among different energy entities in smart grids; and (2) wide-area energy resource management, the use of blockchain to certify and record data communication and control actions in the process of coordinating the control and operations of distributed energy resources.

In the remainder of this chapter, we study four key application scenarios of blockchain in smart grid: energy trading, data management, demand-side management, and emission credit trading. Section 8.2 studies the frameworks that propose the use of blockchain for energy trading. Blockchain has the potential to serve as an infrastructure to manage the data produced by the participating nodes, discussed in Section 8.3. Section 8.4 studies the demand-side management applications of blockchain. Section 8.5 outlines the blockchain applications for emission credit trading. Finally, Section 8.6 concludes the chapter.

8.2 Blockchain for Energy Trading

Traditionally, energy trading is implemented in regional or grid-level energy markets, which are commodity markets supporting energy trading among power generation companies and power distribution companies or power retailers. End users are usually not eligible for participating in energy markets as the energy consumption and/or production capacity of an individual end user is too small to meet the minimum capacity requirement of the participating energy market, also known as a capacity barrier. With the prevalence of DESs, end users that traditionally function as energy consumers have become energy prosumers, which are nodes that can function as energy consumers and producers simultaneously. In such a scenario, end users are naturally more inclined to trade electricity with each other. For instance, when the energy generated by the solar power panel of a smart home exceeds the energy consumption of the home, the users can sell the surplus energy to their neighbors who need energy, potentially at a price cheaper than the price of purchasing electricity from the grid. Then the key questions are how such prosumers can have trusted energy transactions with other participants on the grid, while remaining anonymous to protect their privacy and how these transactions can be supported at a large scale.

Recall from Chapter 3 that blockchain builds a trusted network over untrusted participants without reliance on central authorities and introduces a level of auditability. Thus, it can potentially be a promising infrastructure for the distributed energy trading scenario. In the rest of this section, we outline the details of some distributed energy frameworks. Aitzhan and Svetinovic [5] have designed a decentralized energy trading platform, known as PriWatt, that relies on blockchain as its backbone to facilitate energy trading among energy prosumers. Energy is converted into tokens, and the ownership of the tokens is transferred using an energy trading event that is a transaction digitally signed by the initial owner and is represented as:

$$nVersion \| vinNum \| vin \| voutNum \| vout \| nLockTime$$

where *nVersion* is the hash of the previous transaction in the ledger, *vinNum* and *voutNum* represent the number of inputs and outputs of the transactions (i.e., sources of energy), respectively. *vin* and *vout* are input and output vectors represented by tuples of elements that are used for transferring token ownership from the previous owner to the current owner, and the current owner to the next owner. *nLockTime* represents an expiry time. The transactions are included in the blockchain after nLockTime has passed and can be replaced before this time. PriWatt employs POW as the underlying consensus algorithm due to the high resilience of POW against malicious nodes that may attempt to dominate the blockchain and store fake information in the chain (see Chapter 3). In a centralized trading platform, the centralized authority ensures that both sides of the trade commit to their obligations. The absence of the central authority makes it challenging to ensure that both sides commit to their obligations. To address this challenge, PriWatt uses multisignature transactions where two of three parties must sign the transaction to be considered as valid. The third participant in the transaction is a trusted party by both sides of the transaction. If the consumer receives energy but avoids paying energy price, the trusted party signs the transaction, which will then pay the energy price to the energy producer. In addition to the buyer and seller, the distribution company also participates in the trade to ensure that both sides of the trade commit to their obligations and also serves as a mediator in case of dispute between the participants. To protect against a double-selling attack where a malicious energy producer attempts to sell energy to multiple users, PriWatt locks the energy tokens once the seller and buyer agreed on the energy price. The seller informs the distribution company of the ID of the locked token. The buyers can query the distribution company for the ID of the sold tokens before purchasing energy.

Li et al. [6] developed a blockchain-based framework for energy trading where IoT devices coexist with the distribution system. The framework employs a consortium blockchain where authorized nodes can join the blockchain and selected nodes participate in mining process. The framework consists of IoT devices (e.g., smart meters and energy aggregators). Energy aggregators are devices with sufficient resources to collect and aggregate the energy consumption/generation pattern from the prosumers. The prosumers send their energy demand/load to the energy aggregator. The energy aggregator aggregates the total energy demand/load and attempts to sell or buy energy from the main grid. The energy aggregators determine the energy price based on the balance of energy production and demand in the network. The energy price is paid using an energy coin that is defined as a currency. To reduce the processing and packet overheads and reduce the energy wasted by transmitting energy, it is critical that the energy aggregator can persuade the local energy producers to produce energy demanded by the prosumers in their neighborhood. As an incentive, the energy aggregator rewards the energy producers by payment of energy coins. To reduce the delay associated with payments of the energy price, the framework introduces a credit-based payment method. A centralized trusted bank manages the credits of the nodes and protects against double spending.

The Brooklyn Microgrid (BMG) project [7, 8] is a representative, real-world demonstration of applying blockchain for peer-to-peer energy trading that was launched by LO3 Energy. The project established a local energy market in Brooklyn, New York. The participants are residents located across three power distribution networks under the coverage of BMG. Through BMG, participants can sell the excess of the solar power produced by their rooftop solar panels to their neighbors. BMG includes a virtual community energy market platform that functions as the cyber infrastructure to trade energy and a physical microgrid, which is an electrical power grid that is built to complement the existing grid and to provide backup assistance during power outages. A private blockchain is employed as the core component in the virtual community. In emergency situations, the physical microgrid can decouple from the main microgrid and supply energy to critical facilities (e.g., hospitals and schools) at fixed electricity rates. The residences and businesses can then bid for the physical microgrid's remaining power.

The participants of BMG perform electricity trading following a double-auction mechanism in the virtual community energy market platform. Both electricity sellers and buyers submit bids to the market every 15 minutes. The market then stacks the bids based on the bidding prices and determines the clearance price. If the energy price offered by a buyer is lower than the bidding price, the buyer-requested energy will be provided from additional energy resources (e.g., distribution companies).

The existing blockchain-based solutions for energy trading, including the representative approaches outlined above, suffer from the following challenges:

1. *Lack of privacy:* The participating nodes observe privacy-sensitive information about the user, including energy consumption/generation patterns. A malicious node may attempt to deanonymize a user by linking multiple transactions in the blockchain (see Chapter 7), which compromises the privacy of the user.
2. *Reliance on TTPs:* It is critical to ensure that both sides of a trade commit to their obligations. To address this challenge, most of the existing works rely on the existence of a TTP, which is an entity in the network with a partially centralized trust. The reliance on a TTP raises challenges that are inherent with centralization. Additionally, the TTP can build a virtual profile of the users and thus compromise their privacy.
3. *Blockchain overheads:* Conventional blockchains suffer from packet and processing overhead due to a high-resource-consuming consensus algorithm and requiring broadcasting of all new blocks and transactions to all the participants. Such overheads limit the scalability of the blockchain and increase the cost.

To address these limitations, we developed a secure private blockchain (SPB)-based energy trading platform [9] that proposes a series of mechanisms in the energy account creation stage, negotiation stage, and energy trading. We outline the details of SPB here:

- Energy account creation: In SPB, the energy producers chain all their transactions in a ledger. The generation of a genesis transaction (see Chapter 3) involves burning a specific amount of digital coins by paying them to an unknown address that is considered a registration fee of the energy account. This also protects against Sybil attack (see Section 4.3.3). As outlined in Chapter 7, using the same ledger to chain all transactions of a user may potentially introduce anonymity risks. To address this limitation, an energy producer may create multiple energy accounts to distribute its transactions among multiple ledgers, which enhances the anonymity level of the user. Obviously, creating more energy accounts would lead to an increase in the registration fees and communication costs that highlights the trade-off between privacy and cost, which is analyzed in [9].

  The energy producer advertises its energy in the blockchain by generating and storing an energy transaction that is structured as follows:

$$T\_ID \| P\_T\_ID \| energy\_amount \| energy\_price \| PK^{+} \| Sign \| negotiable$$

  where *T_ID* and *P_T_ID* are the identifier of the current and the previous transaction respectively. energy_amount and energy_price are the amount and price of the energy. PK^+ and sign are populated by the energy producer. Finally, negotiable indicates whether the energy price is negotiable which is outlined in details below.

- Negotiating energy price: SPB enables the energy consumers to negotiate the price of the energy with the producers who are willing to do so. The energy producers may set a negotiation flag in their energy account, which informs the consumers that the energy price is negotiable. To negotiate the energy price, the consumer generates a negotiation transaction that is structured as follows:

$$T\_ID \| energy\_account.PK^{+} \| price \| status$$

  where *T_ID* is the identifier of the negotiation transaction; energy_account.PK^+ is the PK^+ of the energy account with which the energy consumer intends to negotiate; price is the energy price populated by the consumer; and status is a binary flag indicating if the given price is rejected ("0") or accepted ("1") by the producer. The producer can offer another price by setting status as "0" and price as its intended price. In this way, the energy producer and consumer can iteratively exchange the negotiation transaction until they reach an agreement.

Conventional blockchains broadcast all transactions, including the negotiation transactions that introduce significant packet and processing overhead. To address this limitation, SPB proposes a routing mechanism, referred to as the Anonymous Routing Backbone (ARB), that routes the transactions among the participating nodes and thus reduces the overheads. In the ARB, data packets are routed towards nodes with specific PK^+. A group of nodes with higher-resource availability (e.g., a control center in the grid) forms a backbone network and participate in routing the transactions. The backbone nodes form a distributed hash table (DHT). Each backbone node is responsible for managing transactions for the PK^+ of the transaction recipients that start with a range of key-value pairs as shown in Figure 8.2. Other participating nodes, referred to as the regular nodes in the rest of this section, send join messages to the backbone node responsible for managing the transactions corresponding to their PK^+. As the nodes may employ multiple PK^+ as their identity, a node might join multiple backbone nodes (e.g., node 6 in Figure 8.2). The backbone nodes use conventional routing algorithms to route transactions among themselves.

- *Energy trading:* Once the energy producer and consumer agree on the price and amount of energy, they start trading the energy. Recall that it is critical to ensure that both sides of the trade commit to their obligations without relying on TTP. To address this challenge, the SPB introduces the concept of atomic meta-transactions that shares similarities with atomic processes. An atomic process is an indivisible operation that appears to the rest of the system to occur at once without being interrupted. The atomic meta-transactions are the Commitment-to-Pay (CTP) transaction and the Energy Receipt Confirmation (ERC) transaction, which are outlined in detail in the rest of this section. Neither of these

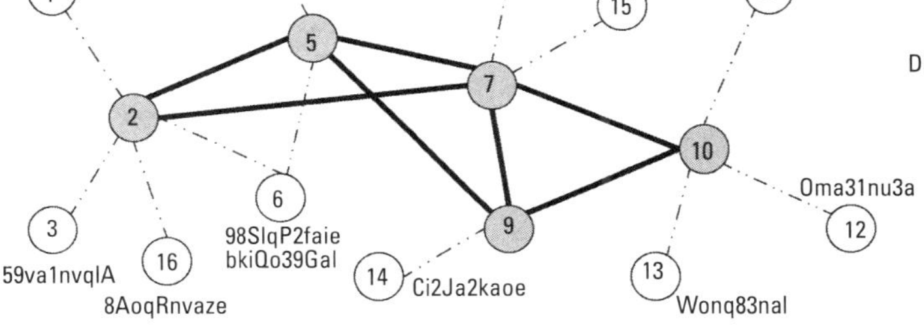

Distributed Hash Table

Node	Key
2	0-9
5	a-k
7	l-z
9	A-K
10	L-Z

Figure 8.2 An overview of the SPB routing algorithm.

transactions is considered valid without being accompanied by the other transaction, which must be created within a particular time period.

The CTP transaction is generated by the energy consumer, representing the consumer's committed payment (agreed in the negotiation stage) to the energy producer. The payment is with pending status and the money is not actually transferred to the producer's account until the ERC transaction (introduced later) is completed. Storing all transactions in the blockchain increases the blockchain database size and the cost for the end user as the user must pay the transaction fee. In the SPB, the CTP is stored in a temporary database known as CTP database and later can be removed. The miners include the hash of their version of the CTP database in the blocks that they generate, which ensures the integrity of the CTP database. When a CTP transaction is broadcast and received by the energy producer, the producer generates the agreed amount of energy. The energy is delivered to the energy consumer, which is recorded by the smart meter of the consumer. The smart meter then generates an ERC transaction to confirm receipt of the energy. By receiving the ERC, the miners check if the corresponding CTP had been previously generated. If so, they mine the ERC in the blockchain, which triggers the payment of the energy price to the energy producer, following the CTP transaction being removed from the CTP database.

In this section, we discussed the blockchain applicability for distributed energy trading and outlined the relevant challenges and requirements. Blockchain is also widely applied to manage the flow of data in the smart grid, which is detailed next.

8.3 Blockchain for Data Management in Smart Grids

A smart grid is a highly distributed computing environment produced by a wide deployment of sensing and measuring devices that include smart meters and phasor measurement units. The data produced by these devices needs to be collected and stored in a trusted and secure manner. Traditionally, data generated by the wide-area sensors and meters are gathered and transmitted to the control center by the Supervisory Control and Data Acquisition (SCADA) system; however, this scheme is vulnerable to attackers against data integrity and a single point of failure. As a trusted distributed database, blockchain has been explored for the management of smart grid data.

In [10], Liang et al. designed a data gathering and protection scheme based on blockchain that reconfigures the centralized SCADA system as a distributed blockchain. The meters are equipped with computing and communication chips, making them capable of packaging the meter readings as blocks that can be stored into the blockchain ledgers through a certain consensus mechanism. Each meter populates a transaction with the plaintext data and the corresponding hash to ensure integrity. The receiving nodes must verify the data and collaboratively reach consensus over the validity of the data that is achieved using a voting mechanism. Each node that receives the transaction verifies the data by hashing the plaintext data and comparing with the hash stored in the transaction. If matched, the node votes on the validity of the data by signing the transaction. A data item is considered to be valid only when the following criterion is satisfied:

$$\frac{K}{N} \geq T$$

where N is the total node number in the network, K is the number of nodes voting for the validity of the data, and T is a threshold value that must be larger than 50%. The data items that pass the data verification are packaged into multiple blocks and are then sequentially inserted into a ledger. Each block consists of multiple data items. The framework has designed two consensus strategies to package data items into blocks: (1) generation of blocks within a fixed time where all data items that are verified within a specific time period are packaged into a block; and (2) generation of blocks within a fixed size where each block contains an equal number of data items. Depending on the network load and the delay that can be tolerated for the transactions to be stored, one of the outlined methods can be employed. In the networks with fewer transactions, generating blocks with a fixed period of time may reduce the delay as the generation of enough transactions to form a block may take longer time. The framework designs two strategies for choosing the miners: (1) the use of prespecified nodes as miners, in which a certain number of nodes are prespecified to be responsible for solving the POW; and (2) the random selection of nodes as miners, in which, for each mining process, miners are randomly selected from all the nodes in the blockchain network.

For the operation of a power grid, the system operator needs to know the regional power load of different geographical areas. The regional load represents the total power consumption of an area, which is calculated by summing up the individual smart meter readings in the area. Normally, smart meter data are transmitted to substations or control centers where Energy Management Systems (EMS) can calculate the regional load based on the individual meter readings. Wang et al. [11] designed a distributed smart meter data aggregation framework to support secure and reliable regional load calculations. The framework is based on blockchain and the Homomorphic Encryption (HE), a technology that allows performing algebraic operations on encrypted data without decrypting them.

The framework is based on a hierarchical blockchain structure that consists of multiple Regional Cluster Blockchains (RC-BCs) and a Wide-Area Blockchain (WA-BC), as shown in Figure 8.3. Smart meters in the same geographical area are grouped as a cluster and managed by an RC-BC. The readings recorded by the individual smart meters are stored in the RC-BC in a similar way as presented by Liang et al. [10]. Each RC-BC is equipped with a cluster gateway that also acts as a node of the WA-BC. The WA-BC consists of multiple cluster gateway nodes that belong to different clusters and the control center node.

The individual meter data readings are encrypted and summed in the cluster gateway by using the HE. The calculated sum of the individual smart meters in a cluster is then encrypted, broadcast, and stored in the WA-BC. The control center node then uses HE again to compute the sum of each cluster's total load. In this way, the control center knows the power load of the whole region without revealing individual meters' plaintext readings. The HE sum of K encrypted smart meter readings is calculated as follows:

$$D = \phi_1 + \phi_2 + \ldots + \phi_K = m_1 + m_2 + \ldots + m_K \tag{8.2}$$

where D is the sum of the smart meter readings, ϕ_1 and m_k ($k \in [1,K]$) are the encrypted value and the plaintext of the kth meter reading, respectively. The details of the Paillier homomorphism procedures can be found in [12].

In summary, this section outlined the key features enabling blockchain to serve as communication framework to exchange data of the energy participants

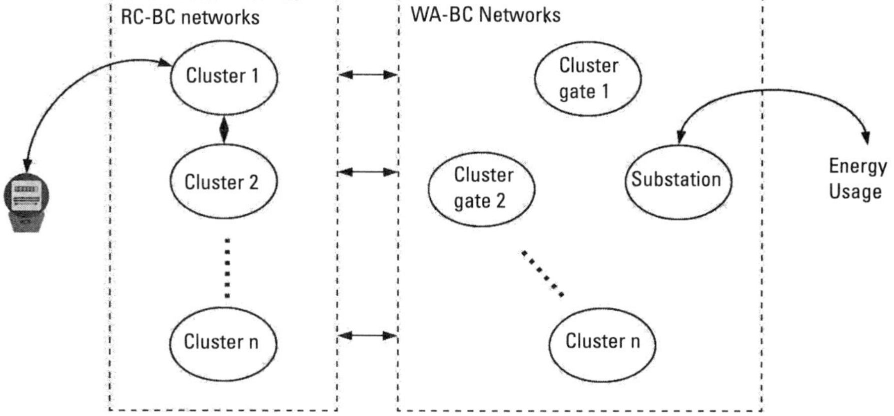

Figure 8.3 An overview of the framework outlined in [11].

in the smart grid. We next study the blockchain applications in the demand-side management.

8.4 Blockchain for Demand-Side Management

In traditional power systems, the power generation is completely driven by the power load. With this approach, the utility schedules power plants to produce the same amount of power to match the demand of power consumers. This one-way communication pattern can have a negative impact on power grids. A typical example is the network peak load problem. On hot summer days, a large number of end users operate air conditioners simultaneously, which leads to very high grid peak power loads. The utility needs to schedule the output of power generation resources and to deliver the power to the load side to match the peak power loads. Such high peak loads could pose challenges to the grid infrastructure and could be harmful to the grid's secure and reliable operations.

In smart grids, the above one-way communication pattern between end users and the utility is transformed into a two-way communication pattern. With this approach, the utility monitors the user's power demand and arranges power outputs to match the demand. The utility can also send pricing and/or incentive signals to end users who, based on these signals, can reshape their power consumption patterns, such as shifting some appliance usages to other time periods or reduce some of their power consumption. This methodology is known as DR [3] in the literature. The word response refers to the fact that end users can reshape their power consumption patterns to respond to the signals sent by the utility. The process of making the pricing/incentive signals and the user-side energy management programs are referred to as demand-side management (DSM) [3].

DSM can be categorized into two classes: pricing-based and incentive-based. In pricing-based DSM, the utility sets up varying electricity price tariffs. The most common one is the Time-of-Use (TOU) tariff, in which electricity prices are different in different time periods of a day. The time-varying prices would encourage end users to shift their appliance usage time to save energy costs. Incentive-based DSM is also known as direct load control (DLC). In the DLC, the utility directly controls some major appliances at the user side in some particular time periods (e.g., temporarily shut down air conditioners in peak load hours). As a subsidy, the utility provides the end user certain incentive rewards (e.g., cash and electricity price discount).

Cioara et al. introduced a near real-time framework to manage the demand response in the smart grid [13], which consists of three main layers: field data aggregation, core backbone platform, and front end. The field data aggregation layer is the first layer where the data is collated from the participating nodes and is streamed to the upper layers. The backbone platform consists of the blockchain as the underlying communication framework where all participants exchange data. Big data analytic engines and the node that runs machine learning algorithms are also part of this layer that processes the data to predict the future energy demand/load. The front-end layer enables the users to access the network through APIs. A smart contract is deployed in the blockchain that collects the energy demand and load and generates regulation signals to manage the load in the network.

In DSM processes, there are several interactions and information exchanges between the end user and the utility, which increase the packet overhead and bandwidth consumption of the participating nodes. Additionally, the exchanged data includes privacy-sensitive information and thus may potentially raise privacy concerns. To address this limitation, in [14], we presented a blockchain-based communication architecture for the DLC. We employed a permissioned blockchain where only authorized nodes by the utility can join the blockchain.

The devices send their transactions periodically to the DSM by generating a demand load (DL) transaction that is structured as follows:

$$ID||Data||DLFlag||Secret$$

where ID is the identifier of the transaction. During the bootstrapping stage, each node generates an ID, a pattern, and a secret value and shares with distribution company (DISCO). To protect anonymity, the node changes the ID for each transaction using a pattern (e.g., adding the previous ID with a constant value). Conceptually, this is similar to changing PK in conventional blockchains. To verify the transactions, DISCO stores all these fields and updates them accordingly once it receives transactions (outlined in detail in the rest of this section). Data is the amount of demand or load for the user. *DLFlag* identifies whether the value in the Data field is demand (DLFlag=0) or load (DLFlag=1). To reduce the overheads on nodes for generating DL translations, a Secret is stored in the last field, which is the hash of three parameters, $H(secret\ value||nonce||Data)$, where $H(x)$ is the hash of x. The secret value is generated by the node during bootstrapping. Nonce is a one-time value used to prevent the replay attack, which conceptually is similar to the nonce employed in the Ethereum transactions [15]. The Data value is included to prevent any node from changing the data.

Each node generates and broadcasts the DL transactions in predefined time intervals. Once received, the DISCO extracts the ID of the transaction generator and locates the stored details (secret value, nonce, and pattern) for that ID. Then the DISCO generates Secret, as outlined above, with received Data and the corresponding nonce and secret value in its own record. If the resulting hash matches with the Secret, the transaction is verified. The DISCO then updates nonce and ID for the node for future transactions. If the hash does not match with the Secret, the transaction is dropped. DISCO creates a Merkle tree of all received DL transactions in a period. Next, it stores the signed root hash of the Merkle tree in the blockchain. Storing only the root hash of the Merkle tree in place of all received DL transactions reduces the processing overhead and memory requirement for managing the blockchain and thus increases its scalability.

In the DLC, the DISCO manipulates the usage of devices on the customer premise to manage the load, based on the received DL transactions. DISCO and the customer would typically sign a contract that allows DISCO to control energy usage in the customer site. The DISCO then installs sensors at the customer site to measure real-time data to prevent any service inconvenience. The utility initializes a contract-signing process by generating and signing a load control transaction. The transaction specifies the number and types of the user-side energy devices that the utility intends to control, together with the intended control duration and control action type. The load control transaction is then broadcast to the user. If the users agree to the terms involved for the load control, they sign the transaction and broadcast it to the blockchain system.

Based on the agreed load control transaction, the utility informs the target device using a load control transaction. After the transaction is received, the device verifies the transaction and performs the control actions specified in the load control transaction.

This section outlined the applicability of blockchain for demand-side management in the smart grid. We studied the existing literature that outlines how the smart grid can benefit from blockchain for demand-side management. Climate change is tightly intertwined with energy systems, particularly through approaches that control carbon emissions. We discuss the blockchain applications in emission credit trading next.

8.5 Blockchain for Emission Credit Trading

The introduction of the carbon credit trading is mainly attributed to address and support climate change issues, mainly caused by greenhouse gas emissions produced from human activities. Carbon dioxide, which is mainly produced from fossil fuel combustion, account for 75% of the global greenhouse gas emissions in 2010 and is estimated to rise at an average rate of 0.6% per year between 2015 and 2040 [16]. Blockchain can be applied to support the trading of carbon emission credit trading, also known as carbon cap trading when applied to carbon dioxide.

Carbon credit trading imposes prices on emission products to encourage industrial firms to reduce emission production. In carbon credit trading, an authority sets an emission limit or a cap to the type and amount of greenhouse gas that the participating enterprises are allowed to produce. At the beginning of the operation period, the tradable credits are created and allocated to each participant. During the operation period, the participants can trade emission credits with each other. Participants with excess credits can sell them to others who need more credits to cover more emissions. At the end of the operation period, all participants are required to follow a certain settlement scheme aimed at penalizing participants that produce more emissions than the amount allowed by their credit.

Khaqqi et al. [17] outlined a blockchain-based platform for supporting the emission trading. The blockchain system interfaces a reputation-based system. The reputation of a participant is calculated by a function of the past emission rates of the participant and the strategies implemented to achieve the emission reduction. Prior to emission trading, a seller or buyer collects bids and offers from other buyers and sellers in the market. The system uses the reputation value of the participants to sort the bids and offers and control the number of bids and offers that the seller or buyer can collect. Participants with a higher reputation receive better bids and offers of access to the market. After a seller and a buyer reach an agreement on the trading emission credits, the transactions are settled by the blockchain system.

In this section, we covered emission credit trading using blockchain, which shares similarities with energy trading. The distributed management and trust offered by the blockchain make it an attractive candidate for such frameworks. However, there has so far been limited work in this area and it remains open for future research and innovation.

8.6 Conclusions

The emergence of distributed energy resources, such as solar panels, and two-way communication methods have revolutionized the traditional grid networks, paving the way for smart grids where a large number of devices collect information about energy communication and generation patterns. The existing smart grid solutions mainly focus on a central authority that manages energy; however, the centralized methods suffer from lack of scalability as the number of nodes and the volume of produced data is constantly increasing. Blockchain has attracted tremendous attention from academia and practitioners to serve as a distributed trusted solution to manage energy in smart grids. In this chapter, we studied four application scenarios including: energy trading, data management, demand-side management, and trading emission credits. We studied a number of frameworks in each of the outlined applications to show the applicability of blockchain.

In this chapter, we have seen how blockchain can enable CPS systems to support the exchange of pooled physical assets in smart grids. In the next chapter, we turn our attention to the use of blockchain for management of autonomous and connected physical assets, namely, smart vehicles, which present their own unique challenges for blockchain in CPS.

References

[1] "BP Statistical Review of World Energy, 2019," https://www.bp.com/content/dam/bp/business-sites/en/global/corporate/pdfs/energy-economics/statistical-review/bp-stats-review-2019-full-report.pdf.

[2] Farhangi, H., "The Path of the Smart Grid," *IEEE Power and Energy Magazine,* Vol. 8, No. 1, 2010, pp. 18–28.

[3] Palensky, P., and D. Dietrich, "Demand Side Management: Demand Response, Intelligent Energy Systems, and Smart Loads," *IEEE Transactions on Industrial Informatics,* Vol. 7, No. 3, 2011, pp. 381–388.

[4] Dong, Z. Y., F. Luo, and G. Liang, "Blockchain: A Secure, Decentralized, Trusted Cyber Infrastructure Solution for Future Energy Systems," *Journal of Modern Power Systems and Clean Energy,* Vol. 6, No. 5, 2018, pp. 958–967.

[5] Aitzhan, N. Z., and D. Svetinovic, "Security and Privacy in Decentralized Energy Trading Through Multi-Signatures, Blockchain and Anonymous Messaging Streams," *IEEE Transactions on Dependable and Secure Computing,* Vol. 15, No. 5, 2018, pp. 840–852.

[6] Li, Z., et al., "Consortium Blockchain for Secure Energy Trading in Industrial Internet of Things," *IEEE Transactions on Industrial Informatics,* Vol. 14, No. 8, 2017, pp. 3690–3700.

[7] Mengelkamp, E., et al., "Designing Microgrid Energy Markets a Case Study: The Brooklyn Microgrid," *Applied Energy,* Vol. 210, 2018, pp. 870–880.

[8] "Brooklyn Microgrid," https://www.brooklyn.energy/.

[9] Dorri, A., et al., "SPB: A Secure and Private Blockchain-Based Solution for Energy Trading," *IEEE Communications Magazine,* Vol. 50, No. 7, 2019, pp. 120–126.

[10] Liang, G., et al., "Distributed Blockchain-Based Data Protection Framework for Modern Power Systems Against Cyber Attacks," *IEEE Transactions on Smart Grid,* Vol. 10, No. 3, 2019, pp. 3162–3173.

[11] Wang, Y., et al., "Distributed Meter Data Aggregation Framework Based on Blockchain and Homomorphic Encryption," *IET Cyber-Physical Systems: Theory and Applications,* Vol. 4, No. 1, 2019, pp. 30–37.

[12] Priya, S., et al., "Paillier Homomorphic Cryptosystem with Poker Shuffling Transformation Based Water Marking Method for the Secured Transmission of Digital Medical Images," *Personal and Ubiquitous Computing,* Vol. 22, No. 5-6, 2018, pp. 1141–1151.

[13] Cioara, T., et al. "Enabling New Technologies for Demand Response Decentralized Validation Using Blockchain," 2018 IEEE International Conference on Environment and Electrical Engineering and 2018 IEEE Industrial and Commercial Power Systems Europe (EEEIC/I&CPS Europe), June 2018, pp. 1–4.

[14] Dorri, A., et al., "Peer-to-Peer Energy Trade: A Distributed Private Energy Trading Platform," *Proc. IEEE International Conference on Blockchain and Cryptocurrency*, 2019.

[15] Wood, G., "Ethereum: A Secure Decentralised Generalised Transaction Ledger," *Ethereum Project Yellow Paper*, No. 151, 2014, pp. 1–32.

[16] "Global Greenhouse Gas Emissions Data," http://www.epa.gov/ghgemissions/global-greenhouse-gas-emissions-data.

[17] Khaqqi, K. N., et al., "Incorporating Seller/Buyer Reputation-Based System in Blockchain-Enabled Emission Trading Application," *Applied Energy*, Vol. 209, 2018, pp. 8–19.

9

Blockchain Applications in Smart Vehicles

Chuka Oham
UNSW, Sydney, Australia

9.1 Introduction

In the previous chapter, we explored how blockchain can support trustworthy, anonymous, and scalable energy management in smart grids, where the unique feature of the tradeable entity (i.e., electrons) is that they are pooled, and their flow must be monitored and managed by the CPS. In this chapter, we turn our attention to an application area where the core CPS entities involved are autonomous and connected platforms, namely, smart vehicles. These vehicles include a complex network of sensors, actuators, and transceivers that combine to provide autonomy features as well as connectivity. Smart vehicles are intrinsically highly mobile, which makes it challenging to keep track of their location, operational state, and liability in the case of accidents. Unlike smart meters in the smart grid, which are generally physically static while observing a dynamic physical process, smart vehicles are themselves highly dynamic as they monitor equally dynamic physical processes, represented by their surrounding context. These features are particularly challenging for smart vehicle traceability. While some technological solutions, such as GPS receivers, can provide some of these functionalities, the generated data can typically expose sensitive information about the vehicle owner. It is also quite difficult to ascertain the data's authenticity, particularly when certain stakeholders may have an interest in altering the data. This chapter explores these challenges of smart vehicles and their associated solutions that are supported by blockchain. First, we provide a more detailed overview of smart vehicles as an application area of CPS.

Smart vehicles are instrumented with a wide array of sensors, communication technologies and specialised and complex computers called electronic control units (ECUs) to provide advanced vehicle functionalities, facilitate driving decisions, enable better perception of the road, and avert road hazards. In this chapter, we focus on smart vehicles capable of independent decisions given that compared to the state-of-the-art vehicles, these vehicles possess multiple attack vectors and present a higher risk for the owner, the manufacturer, and the in-vehicle passenger [1].

The growing adoption of smart vehicles and the increasing volume of data collected by these vehicles introduce security, safety, and privacy challenges. The authors in [2, 3] demonstrated the possibility of remote exploitation of a smart vehicle which allowed a remote entity to gain full control of critical vehicle functionality and bring the vehicle to a halt, which raises serious safety concerns for the vehicle occupants. A smart vehicle collects large volumes of privacy-sensitive information of the vehicle owners, such as their location. Given the potential for external communication and remote exploitation of the vehicle as demonstrated in [2], such information could be used to track vehicle owners and compromise their privacy. In smart vehicles, a broad range of sensors and devices is installed that exchange data to facilitate independent decisions; however, the potential for remote exploitation necessitates the verification of the integrity of the sensors and the credibility of the data that it generates. Thus, security solutions for smart vehicles should also offer sensor integrity verification and data veracity to detect when the vehicle is compromised and prevent the exchange of malicious data with potential life-threatening outcomes for an in-vehicle passenger or vehicle owner.

To address these challenges, researchers and practitioners have conducted extensive research in this area. In [4], an authentication mechanism was introduced that enhances the security of intervehicular communications and the integrity of the received messages. Intervehicular communications refer to communications between vehicles (i.e., vehicle-to-vehicle (V2V) and between vehicles and roadside infrastructures (V2I)). The integration of these entities for information exchange constitutes the automotive or vehicular network where such information exchange optimizes road safety. In [5], the authors introduced a framework to identify unauthorized access to the sensors and devices in the vehicle and thus protect against malicious nodes that may intend to interrupt the vehicle functions or access its data. However, these existing frameworks make use of centralized architectures to manage vehicular security and privacy, which poses scalability and availability issues and introduces a single point of failure.

Due to its salient features including immutability, anonymity, security, and decentralization (see Chapter 3), blockchain can potentially serve as a framework to address the outlined challenges in smart vehicles. In this chapter, we study recent work on blockchain-based frameworks for smart vehicles. In Section 9.2, we review the state-of-the-art approaches for securing automotive networks and we highlight their limitations. We present a new blockchain-based solution for the automotive ecosystem in Sections 9.3 and 9.4 that conclude the chapter.

9.2 State-of-the-Art Solutions

The integration of blockchain with automotive technology has recently gained significant attention owing to the aforementioned features, with several solutions being discussed for securing smart vehicles using blockchain. We classify these works as automotive network security (see Section 9.2.1), trust and reputation management (see Section 9.2.2), privacy (see Section 9.2.3), and vehicular forensics (see Section 9.2.4). Automotive network security solutions address specific issues for automotive networks such as secure interaction, key management, and misbehavior detection, while works on trust and reputation management focus on establishing the credibility of vehicles or credibility of message exchanged by vehicles while operational. Privacy-focused works ensure that sensitive and confidential information about a vehicle owner such as location and identity is not disclosed without the consent of the owner. Finally, vehicular forensics address the issues of gathering requisite evidence for establishing liability in the event of an accident. Table 9.1 summarizes the studied literature and outlines their security properties and limitations.

9.2.1 Automotive Network Security

As outlined earlier in Chapter 3, the communications between blockchain participants are secured using asymmetric encryption. The hash of the exchanged data is signed and included in the transactions, which ensures data integrity. These features along with the ability to have an audit trail of all transactions in the blockchain make it an attractive solution to secure communications and data exchanges in smart vehicles including intra-vehicle communication between in-vehicle components such as ECUs and intervehicular communications with other smart vehicles and roadside units installed and managed by the transport authority of a state.

In [6], we presented a security architecture for automotive applications using blockchain. The blockchain functions as the underlying communication framework that connects vehicle manufacturers, drivers, software update providers, and roadside infrastructures. To reduce the associated packet and

Table 9.1
Practical Limitations of Existing Works

Classification of Earlier Works	Existing Work	Achieved Security Properties	Remarks (Practical Limitations)
Automotive network security	Dorri et al. [6]	Authentication, authorization, privacy, and availability	Approach does not consider the security of a vehicle while operational and the security of vehicular networks as their approach is mostly feasible when a vehicle is static
	Heijden et al. [7]	Authentication, authorization, traceability, and non-repudiation	Vulnerable to collusion attacks and given the ephemeral nature of vehicular networks, reliance on vehicles to provide consensus on behavior of vehicles in the network is impractical
Trust and reputation management	CreditCoin [8]	Privacy and authentication	Vulnerable to tactical attacks such as collusion and propagation of falsified information
	Yang et al. [21]	Authentication and authorization	There exists a high cost to establish a blockchain in resource constrained vehicles using computationally intensive POW consensus algorithm; also, this proposal is vulnerable to propagation of falsified information in the network
Privacy	OnionChain [9]	Privacy, authentication, and traceability	Vulnerable to propagation of falsified information in the network
	Malik et al. [10]	Authentication and privacy	Does not assure the integrity of ECUs and is therefore vulnerable to propagation of false messages and ECU tampering
Vehicular forensics	Cebe et al. [11]	Authentication, availability, non-repudiation, traceability, and authorization	Vulnerable to selective data sharing and does not consider the veracity of data which reflect if ECUs are compromised
	Oham et al. [12]	Authentication, availability, non-repudiation, traceability, and authorization	Does not consider data veracity and vulnerable to collusion attacks

processing overheads, we cluster the network and only the cluster heads (CH) manage the blockchain (which is conceptually similar to LSB as outlined in Chapter 4). Multiple applications could leverage this framework including remote software update, auto insurance, electric vehicle charging, and car-sharing services. The remote software vehicle update involves the software provider, the vehicle manufacturer, the vehicle, and cloud storage. The software provider first creates a new update for the vehicle and uploads it to the cloud storage. The software provider then generates a multisig transaction, which requires two signatures, including its own signature and that of the vehicle manufacturer, to be considered as valid. This is to enable the vehicle manufacturer to verify the content of the update by examining and testing the software code before signing and broadcasting the transaction in the blockchain. The hash of the uploaded data is signed by the software provider and included in the multisig transaction. The CHs notify all the smart vehicles of the new available update. This includes all models of a specific brand produced by the manufacturer (e.g., all Honda Civic 2020 models). These vehicles can download the update from the cloud server and compare the hash of the downloaded file with the hash stored in the blockchain. If the hashes match, the software is updated by the vehicles.

In a separate line of research, Lei et al. [13] designed a blockchain-based dynamic key management framework for secure intervehicular communications. The security of intervehicular communications relies on the safety of the messages transmitted between interacting entities. To achieve this, vehicles within the coverage region of a roadside unit form a group and communicate securely by encrypting messages with a group key, which ensures that only members of the group are privy to messages communicated within the group. While this approach achieves the desired security for secure communication, it introduces a problem of key management including key distribution and key updates, most especially when a group member leaves the group to achieve forward and backward secrecy. The authors integrated RSUs as security managers to transmit and verify vehicle keys while blockchain enables decentralized key management across regions. For this, security managers pick up beacons transmitted by vehicles indicating a request to participate in vehicular communication in the region. These requests are collected as transactions and further gathered into a candidate block. This block is then broadcasted in the security manager's network (SMN) for verification using the POW protocol by Bitcoin [14]. The mined block (MB) is then returned to the SMN after POW is achieved and the new region retrieves the joining information of incoming vehicles from MB, thus reducing the time overhead for key verification.

While the proposal in [13] achieves secure intervehicular communications (IVC), it has to keep track of actions that exploit IVC and potentially threaten the safety of vehicles and passengers. To achieve this, Heijden et al. [7] designed Blackchain, a blockchain-based accountable revocation mechanism, which identifies malicious actions executed by vehicles and revokes their IVC capabilities. This proposal integrates vehicles, RSUs, and misbehavior authorities (MA) in a hierarchical blockchain to facilitate efficient revocation of malicious vehicles and identification of compromised MAs responsible for revoking the communication rights of vehicles. To detect malicious vehicles, vehicles on a particular trajectory form clusters and agree on a cluster state and state changes that they forward to an RSU. The cluster state reflects cluster participants and cluster communications. Furthermore, it allows cluster participants to verify the behavior of other cluster participants by verifying the authenticity of their messages. The cluster state forwarded to an RSU is then propagated to other RSUs in a given region, and using the byzantine fault tolerance protocol, they achieve consensus on data received from clusters and forward the resulting consensus data to MAs. The MAs use the information provided by cluster participants to perform revocation actions. Given that consensus reached by RSUs rely significantly on the input of cluster participants, the use of a permissionless blockchain allows consensus data to be publicly verified and the veracity of the decisions by MAs could be ascertained.

9.2.2 Trust and Reputation Management

Smart vehicles share data with other vehicles and RSIs in their vicinity which facilitates the propagation of traffic and/or road information. To prevent against malicious vehicles that may inject fake information (e.g., fake traffic alert), it is critical for the vehicles to establish trust on other vehicles. However, due to high mobility of the vehicles, the neighbors of a vehicle will change over time, which complicates trust and reputation management. Distributed trust and reputation frameworks, as outlined in Section 2.4, facilitate trust establishment among the vehicles. However, such systems suffer from lack of historical information about the behavior of the users and compromise the privacy of the vehicle owner as the identity of the vehicle must be revealed.

To address the aforementioned challenges, blockchain-based trust and reputation management in the context of smart vehicles has received significant attention. As outlined earlier in Chapter 6, blockchain transactions are stored permanently on a public immutable ledger, which makes it an attractive solution for trust management. Participants in the vehicular ecosystem can maintain trust values for their counterparts for guiding data analysis and decisions. The reputation of the participants is linked to the history of actions of the participating nodes, which can serve as a metric to identify malicious activities and isolate them from the network.

Madhusudan et al. [15] presented an Intelligent Vehicle-Trust Point (IV-TP) mechanism for intervehicle communication using blockchain. IV-TP is a unique identifier for vehicles issued by an authorized entity such as a transport authority to facilitate trust management and the identification of illegal actions exhibited by vehicles. IV-TP is incremented when actions executed by vehicles are considered benign and decremented otherwise. A higher IV-TP reflects a vehicle that is highly trustworthy. To validate actions exhibited by vehicles, vehicles form clusters to enable group communication and provide their IV-TP during communication to build trust in the communication network. When data is communicated by any vehicle in the cluster, the group members verify and validate the message generated by such vehicle. For the message to be considered valid, a proof-of-driving consensus protocol is utilized to ensure that a message disseminated in the network is considered successfully verified only when 50% of vehicles in the group can ascertain the legitimacy of the message generator and successfully decrypt the message content. When this occurs, the IV-TP of a vehicle is incremented, and data is stored in the blockchain. Similar to [15], the authors in [16] designed a blockchain-based credibility assessment framework for vehicular networks for secure data sharing. Vehicles moving in a similar trajectory form a cluster where a blockchain is maintained in each cluster that contains the reputation information for the vehicles in the cluster. Prior to participating in vehicular communications, vehicles are registered by a trusted authority. The trusted authority issues communication credentials, that is, public key (PK^+) and private key (PK^-) pairs (for authenticated communications) to vehicles as well as a sensing certificate based on their sensing abilities. The sensing certificates are used to distinguish vehicles based on the quality of data they generate. In this work, data transmitted by a cluster member is only accepted if its reputation value is higher than a preset threshold. The transactions are grouped in the form of blocks and are stored in the blockchain by miners. The cluster member that first finds a hash were the following is met can store its block to the blockchain:

$$H(Identity; curr\_time; PreviousHash) < C$$

where C is the hash threshold of the sensing certificate issued by the trusted authority. Upon winning the election, the miner packs its computed reputation ratings into a block and broadcasts it to cluster members. Cluster members verify if the miner satisfies the election scheme, if the signature on the reputation ratings in the block generated by the miner can be validated, and if the reputation ratings in the block do not conflict with the ratings of the cluster members. If these checks pass, the block is accepted by cluster members.

A similar work by Javaid et al. [17] introduced DrivMan, a blockchain-enabled trust management and data-sharing security solution for vehicular ad-hoc networks (VANET), while also considering privacy. Specifically, DrivMan is able to prove the reliability and trustworthiness of data exchanged in VANET while also preserving the privacy of vehicles exchanging information. Secure communication is achieved using PK^+ infrastructure (PKI) where each vehicle is assigned a pair of PK^+/PK^-. Privacy for interacting entities is also achieved by exploiting the features of PKI for anonymous communication and to remove the linkability of PKIs to actual identities of the vehicles in order to safeguard and secure the vehicle against malicious adversaries. A physical unclonable function assigns unique identifiers to each vehicle. The physical unclonable function is a hardware security primitive that is characterized by a challenge and response pair. Each physical unclonable function produces a unique response when it receives a challenge. If a similar challenge is served as input to different physical unclonable functions, the challenge produces different responses. Two smart contracts, namely, SC1 and SC2, are employed to ensure data integrity and data provenance. SC1 is a public contract that interacts with roadside units and ensures that data generated by smart vehicles is coming from legitimate vehicles to establish data provenance. SC2 is a private contract invoked only by SC1 and is responsible for the storage and retrieval of data from the blockchain.

While the works in [15, 16] focus on ensuring authenticated and trusted communications between vehicles, they have assumed that vehicles would willingly contribute to verifying the legitimacy of vehicles or the credibility of messages exchanged between vehicles; however, such assumptions are not guaranteed particularly in the absence of any clear incentives. Thus, there is a need for an incentivization mechanism to motivate vehicles to contribute to securing vehicular networks. The authors in [8] introduced CreditCoin, a privacy-preserving, blockchain-based incentive announcement and reputation management scheme for smart vehicles. Their proposal is based on threshold authentication where a number of vehicles agree on a message generated by a vehicle and forward the message to a nearby roadside unit. A vehicle that witnesses an accident (say, V) must collaborate with other vehicles in the close vicinity to increase the trustworthiness of its message. V generates a request to other witnesses (p), asking them to confirm the message that contains the accident information and is produced by V. Three types of messages are exchanged between V and the witnesses. The first type, called the request message, is to ask witnesses to agree to the message and sign it. The second is the reply message, which willing witnesses send to V to confirm their participation, and the final message is the aggregated message from V to other witnesses to verify the validity of the aggregated messages. Upon receiving ($p - 1$) replies from the vicinity vehicles, V forwards the message with p confirmations including its generated data to other vehicles in the same trajectory so they can replan their route. Given that witnesses usually lack the motivation to collaborate and provide information, the authors also introduce an incentivization mechanism to enable witnesses share traffic information. To this end, every vehicle has a credit account that contains reputation points called coins. To make an announcement or transmit a message, a vehicle pays coins as incentives to other vehicles and other vehicles who reply to the message earn an amount of coins.

Hitherto, works discussed have focused on ascertaining the legitimacy of vehicles and building the trust for vehicles to achieve secure data exchange; however, clustering for achieving trust for intervehicular communication does not consider the ephemeral nature of the interactions between vehicles and high mobility that characterizes the vehicular communication network and makes it difficult for vehicles to remain in communication for an extended amount of time such as the time required to achieve consensus in [16]. Furthermore, these proposals increase the processing overhead for vehicles for achieving consensus and making reputation decisions.

We have seen so far how blockchain has been proposed as a solution for vehicular security and trust. Most of the works that have focused on applications of blockchain in vehicular settings have lacked a focus on privacy, which has necessitated the development of privacy-preserving algorithms for smart vehicles with blockchain. We discuss these next.

9.2.3 Privacy

Blockchain anonymity (see Section 9.2.7) makes it an attractive solution to enhance the level of privacy for the vehicles while sharing data. The works discussed in this section target vehicle privacy for intervehicular communications via mutual authentication to achieve legitimate communication, layered encryption to prevent eavesdropping, private blockchains, and pseudonyms to restrict access to information and preserve privacy of interacting vehicles.

Shahid et al. [18] introduced a secure, trust-based architecture that utilizes blockchain to increase security and privacy in smart vehicles and prevent medium access control (MAC) layer attacks such as data falsification and modification attacks, impersonation attacks, replay attacks, and a denial-of-service (DOS) attack. These attacks could impact network security and privacy and so threaten the security of data exchanged within the vehicular communication network between network entities and could lead to fatal situations on the road. Therefore, this architecture utilizes blockchain to guarantee the trustworthiness of interacting entities in the vehicular network and increase the privacy of vehicles. These entities interact in a public blockchain, which serves as ground truth for other entities in the network.

The framework contains four blockchains: the certificate blockchain (CertBC), message blockchain (MesBC), revocation blockchain (RevBC), and trust blockchain (TrustBC). CertBC contains the communication credentials of all entities in the vehicular networks. A participant A that intends to communicate with another participant B looks up CertBC to validate B's legitimacy. MesBC stores all similar messages (i.e., transactions) or events reported by entities in the network. RevBC stores the PK^+ of vehicles whose malicious actions have been identified. If the PK^+ of a vehicle exists in the RevBC, its corresponding messages are discarded. TrustBC is utilized to store the trust values for interacting entities. Network entities interact with roadside units to obtain the trust values from TrustBC to make decisions on accepting or rejecting messages from an entity. These blockchains are managed by trusted entities such as law enforcement agencies (LEA) and certificate-issuing authorities. This is to facilitate legal utilization of data for dispute resolution.

The framework contains six phases: (1) system initialization, (2) authentication, (3) message rating generation, (4) trust calculation, (5) miner election and block generation, and (6) distributed consensus. In the system initialization phase, the identities of interacting entities (i.e., vehicles) are validated and a certificate is issued for verifiable communication. The authentication phase enables two communicating entities to mutually verify their identities. To preserve the privacy of vehicles, a vehicle has to change its certificate frequently; otherwise, its certificate expires and is placed in the RevBC. The message generation phase focuses on evaluating the trustworthiness of messages exchanged in the communication network and provides a trust rating based on the evaluation. The fourth phase is the trust value calculation, which evaluates the credibility of entities in the network. In miner election, block generation, and distributed consensus phases, new blocks are added to the blockchain.

A different take on a blockchain-based vehicular privacy approach is presented in [9], where the authors propose OnionChain, for blockchain-based applications and extensively discussed its application in vehicular networks. The inspiration behind their research was based on onion routing, which is designed for anonymous communication. Onion routing utilizes a set of onion routers to encrypt and relay packets between source and destination and is resistant to eavesdropping and traffic analysis via packet encryption thereby achieving privacy. OnionChain employs onion routing on top of blockchain and enables conditional packet decryption. For vehicular networks, OnionChain can provide privacy and identify malicious actions exhibited by compromised/malicious vehicles when an attack occurs. To achieve privacy and traceability, OnionChain offers three key protocols: registration, message transmitting, and identity disclosure. Registration protocol is used to register vehicles to the blockchain network and ensures that only legitimate vehicles are able to communicate in the blockchain. Data communicated between these entities are secured by encryption using negotiated keys. For example, a vehicle intending to send a message randomly to three vehicles negotiates communication keys with each of them. Next, the source vehicle adds three layers of encryption to the message and forwards it to the first vehicle. Upon receiving the message, the first vehicle decrypts the first layer of encryption, finds the next destination of the message and forwards accordingly, and so on, which achieves the desired anonymity.

The work in [18] achieves privacy by allowing authenticated and verifiable communication between two communication entities, while the work in [9] achieved privacy yet verifiable communication via three-layer encryption. While both approaches ensure that communications between interacting entities cannot be eavesdropped and that only interacting entities are privy to exchanged information, they do not fully preserve privacy as communicating parties are privy to sensitive information exchanged during a communication round.

To address this, Malik et al. [10] presented a blockchain-based framework that utilizes a private blockchain to avoid exposure to untrusted entities and enable selective access to the blockchain and utilize pseudonyms for privacy preservation between interacting entities. In the private blockchain, the certification authorities (CA) and revocation authorities (RA) are responsible for managing the blockchain and have read-and-write blockchain access. RSUs have read access, and smart vehicles have no access to the blockchain. Also, this proposal ensures traceability in the case of suspected malicious behavior. To achieve traceability, the proposal utilizes a hash map and a pointer to the blockchain ledger. Vehicles initially register to be a part of the communication network. During registration, vehicles present their vehicle identifier (Vid) to the CA and are issued a pseudo identifier (Pid) and a PK^+ and PK^- key pair. The CA maintains a hash map in the blockchain which is then used to match the *Pid* to the *Vid.* To achieve privacy for vehicles, communication is enabled in the network using pseudonyms. Furthermore, for mutual authentication, when a smart vehicle approaches the area of coverage of a roadside unit, it authenticates itself with the roadside unit and becomes a member of the group of vehicles in the radio range of the roadside unit. While in the radio range of the roadside unit, the smart vehicle forms a message that includes its hash pointer representing its address in the blockchain and encrypts the message with PK^+ of the roadside unit. When the roadside unit receives the message, it decrypts it first and confirms the existence of the vehicle's record in the blockchain. Once confirmed, the roadside unit sends a challenge to the smart vehicle encrypted with the vehicle's PK^+ and waits for the response of the vehicle. If the vehicle decrypts the challenge message successfully and sends the response, it is authenticated by the roadside unit. The roadside unit then provides the vehicle with a group key to facilitate communication with other group members and the roadside units, request data and receive critical and emergency notification from roadside units. The use of pseudonyms only partially protects privacy, as participants are still vulnerable to linking attacks, where multiple transactions may be linked to the same pseudonym, which helps to reveal its real-world identity.

This section has covered the use of blockchain to support privacy in vehicular networks, so that vehicle data is protected from malicious entities. As vehicular networks evolve towards greater connectivity and autonomy, there is an emerging need to have verifiable records of vehicular events for insurance or liability attribution and to share this data as needed. We expand on data trusted data sharing for vehicular forensics next.

9.2.4 Vehicular Forensics

The ability of smart vehicles to contribute to driving decisions disrupts the existing auto insurance model as multiple parties might be liable for accidents (e.g., the driver or the devices). This disruption introduced by independent decision making necessitates the consideration of external entities with vehicle access such as the vehicle manufacturer and service technicians for liability attribution and therefore complicates liability attribution. Furthermore, the potential attribution of blame to vehicle manufacturers and service technicians necessitates the consideration of securing the vehicles against likely malicious actions of these entities to evade liability. Hitherto, securing vehicular networks focuses on mitigating the impacts of external entities and loosely focused on entities such as vehicle manufacturers and service technicians with unrestricted access to the vehicle. Recent works that address these challenges have relied on the integrity, auditability, and immutability features of blockchain to secure smart vehicles against internal adversaries by providing a common and reliable record of events, as we discuss below.

Cebe et al. [11] introduced Block4Forensic (B4F), a blockchain-based solution for forensics that integrates concerned entities including the vehicles, manufacturer, auto insurance company, service technician and the law enforcement agencies into a common blockchain network for liability attribution purposes. Their proposal facilitates the provision of necessary data needed for settling disputes. In this proposal, vehicles transmit accident related data to the blockchain network. The transmitted data are digitally signed to ensure they are authentic. The manufacturer, insurance company, and service technician are responsible for managing the blockchain network. An entity is randomly selected from these managers to propose blocks based on transmitted accident-related data and the blockchain managers run byzantine agreement to reach consensus on a proposed block. Law enforcement agencies are monitor entities that do not participate in the validation process but maintain a replica of the shared ledger for dispute settlement. While this proposal facilitates the provision of necessary data needed for settling disputes, it does not satisfy the conditions for liability attribution [12]. The conditions are to attribute blame to the vehicle owners for neglecting instructions of vehicle manufacturers, such as executing a software update request to optimize the vehicle's functionality, and to attribute blame for the service technicians and vehicle manufacturer for traceable service defects and product defects. To address this problem, we designed a blockchain-enabled security solution for vehicular forensics also known as blockchain based framework for auto-insurance claims and adjudication (BFICA) [12] as an overlay of B4F to satisfy all conditions necessary for liability attribution. BFICA utilizes a proof of interaction and execution mechanism to identify cases of negligence. The proof of interaction is a multisig transaction whereby instructions generated by a vehicle manufacturer, for example, are acknowledged when a vehicle owner appends its signature to the transaction. BFICA relies on data from neighboring vehicles to establish data consistency and ensures that data from a vehicle in the event of an accident is not altered. Furthermore, to ensure that likely liable entities cannot alter data in the block containing evidence, BFICA introduces a lightweight dynamic consensus protocol that preserves the security of data in a block before it is appended in the blockchain.

In summary, the existing works that rely on blockchain for automotive networks have focused on the network's security, trust and reputation, privacy, or forensics. However, so far there has been ittle attention on the vulnerability of the in-vehicle network of electronic control units, which, as we find in the next section, constitutes a significant point of entry for malicious adversaries and source of high-impact threats. Therefore, it is critical to for automotive security to determine how to monitor the internal controllers of smart vehicles and to quickly identify a compromised vehicle so as to prevent the execution of rogue actions. In the following sections, we explore a solution to this challenge.

9.3 Toward New Blockchain Security Solution for Smart Vehicles

In this section, we discuss a novel framework that attempts to address the limitations of the existing blockchain frameworks. To create a security solution that addresses the vulnerabilities of the in-vehicle network, it is pertinent to define security objectives for securing smart vehicles. The objectives would elucidate requisite measures to assure the security of the vehicles. Based on the defined objectives, we highlight security properties needed to achieve the objectives.

9.3.1 Security Objective

Security objectives are defined to specify necessary security measures needed to assure the security of smart vehicles. The general objectives for securing smart vehicles are:

- Securing vehicular communications including internal and external communications;
- Prohibiting communication of a compromised vehicle;
- Communicating entities should not be able to deny their actions.

To achieve these objectives, we define the following security properties:

- *Integrity:* This requires that message generated by a vehicle should not be altered. This implies that the integrity of data-producing sensors of a vehicle should be periodically ascertained to assure the credibility of data that it generates.

- *Availability:* This requires that, at all times, authorized entities should never be denied access to requisite services.
- *Privacy:* Given the large volume of personal information generated by a smart vehicle, such information should be protected from unauthorized access.
- *Authentication:* This requires that communicating entities' identities can be confirmed as authentic.
- *Authorization:* This requires that only entities with necessary privileges have access to certain information.
- *Non-repudiation:* This requires the implementation of an audit trail to keep track of executed transactions.
- *Traceability:* This requires that actions executed an entity can be traced to that entity.

Having defined the security requirements needed to achieve the security objectives, we identify the security properties violated for threats identified in the risk analysis study in [1]. We refer readers to this reference for a more comprehensive discussion on threats and their classification.

Table 9.2 outlines the threats identified for smart vehicles (second column), the classification of the threats according to severity (third column), and the security property (fourth column). The classification of risks (third column) reveals the severity of the threats for intervehicular communications and for smart vehicles. The risk assigns threats that are likely but not fatal as major, while threats classified as critical are both likely and have fatal consequences for vehicular communications and appropriate countermeasures must be developed for them. In the next section, we present a blockchain-enabled countermeasure for smart vehicles and discuss how we achieve identified security properties and mitigate identified threats.

9.3.2 Blockchain-Enabled Countermeasure for Smart Vehicles

In this section, we outline details of the framework. While earlier efforts have focused on providing security solutions for automotive networks, trust and reputation management, privacy, and forensics, they have not considered the integrity of the in-vehicle components such as data-producing sensors, which, as discussed by the authors in [1], are a significant source of critical threats. Thus, our contribution to securing smart vehicles is to present a security solution that

Table 9.2
Violated Security Properties for Identified Threats in [1]

Number	Threats	Risk	Violated Security Property
1	Eavesdropping and traffic analysis	Major	Authorization
2	Location tracking	Major	Privacy
3	Message injection	Critical	Integrity
4	Spoofing	Critical	Integrity
5	Replay	Critical	Authentication
6	Flooding and spamming	Major	Availability
7	Black hole	Critical	Availability
8	Jamming	Critical	Availability
9	Illusion	Critical	Integrity
10	Network intrusion	Critical	Authentication
11	Sybil	Critical	Authentication
12	Message suppression	Major	Integrity
13	Malware	Critical	Integrity
14	Masquerade	Major	Authentication
15	Timing	Critical	Integrity
16	Backdoor	Critical	Traceability
17	Repudiation	Low	Non-repudiation

ascertains the integrity of data-producing sensors of smart vehicles given that to successfully execute a malicious action, an in-vehicle component would be compromised. In addition, we introduce a proof-of-legitimacy mechanism that establishes the validity of interacting entities to ensure that intervehicular communications occur only between legitimate vehicles.

In this section, we describe our blockchain architecture and components including interacting entities and their roles and data exchanged between interacting entities. Our architecture is based on the permissioned blockchain to allow only trusted entities manage the record of vehicles in the blockchain network. This is to ensure that untrusted entities are not privy to sensitive information generated by a vehicle. Figure 9.1 describes the interactions between entities in our security architecture.

For blockchain communications, we utilize the existing public key infrastructure (PKI) such as a CA to issue unique digital identities to entities in our architecture. The certificate or verification component of the CA is stored in the blockchain and used to authenticate and verify transactions. In the following, we describe the role of entities in our framework and discuss the communication exchanges between entities.

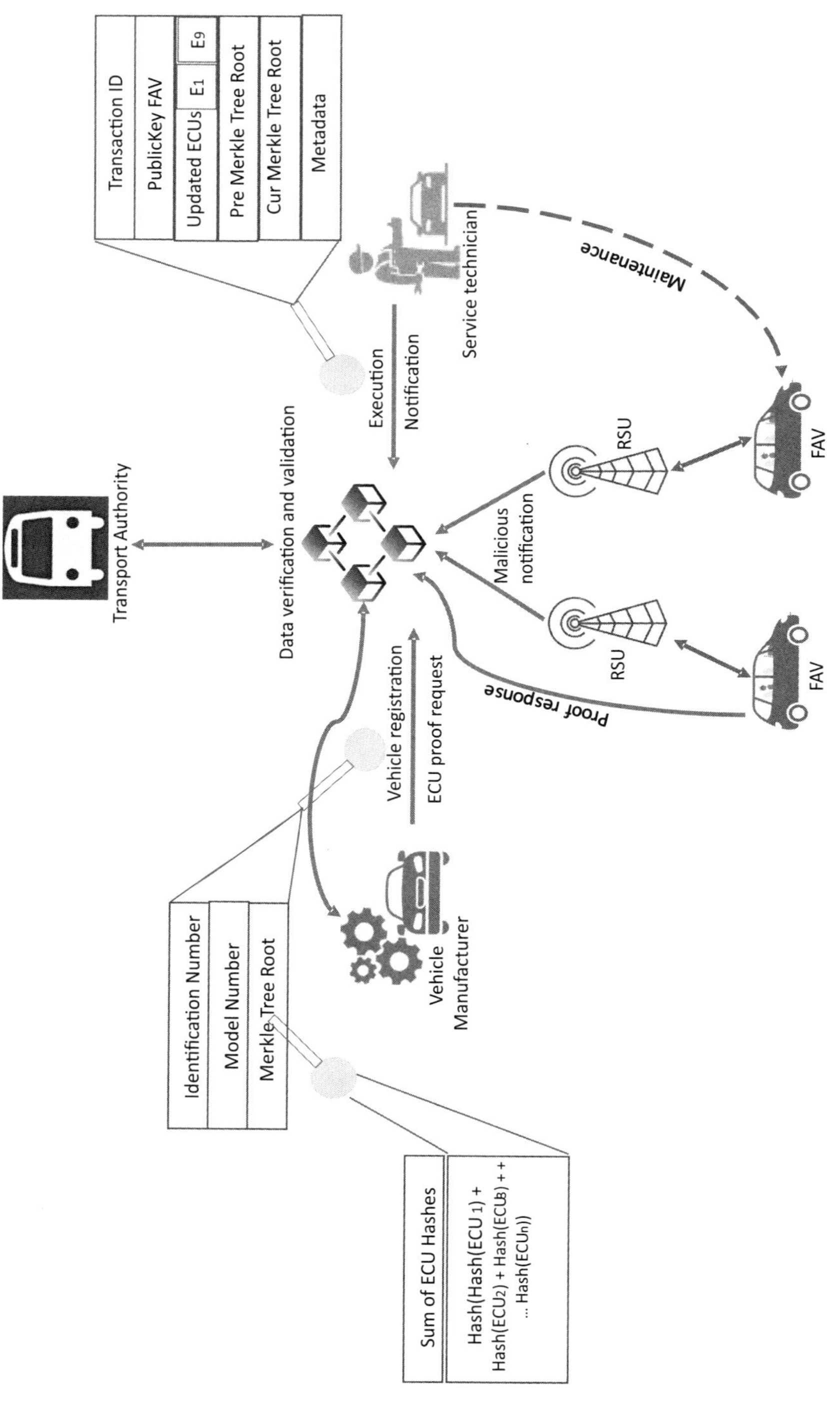

Figure 9.1 Blockchain architecture for securing smart vehicles.

9.3.2.1 Entities

In this section, we describe the roles of interacting entities in our architecture. Entities considered in our architecture include entities considered in our target of evaluation, that is, the roadside units (RSUs), the vehicle, and human assets. Human assets considered in our architecture are entities that have physical and remote access to the vehicle including the vehicle owner, service technician, and vehicle manufacturer. We also include the transport authority, which is responsible for managing the road networks and installations RSUs to monitor road traffic. These entities are integrated in a blockchain network where data exchanges contribute to securing communicating entities and vehicular communications. In Section 9.3.2.2, we describe the data exchanged between interacting entities.

- *Smart vehicle (autonomous vehicle (AV)):* The vehicle contributes to vehicular communication by generating data about road conditions and communicating this data to other vehicles and RSUs. To ensure valid communication, a vehicle must prove its legitimacy once it connects to an RSU. Also, it receives update instructions from its manufacturer for optimized performance.
- *Vehicle manufacturer:* The vehicle manufacturer periodically receives sensor data information from an AV for diagnostics and maintenance. It also provides over-the-air (OTA) software updates to an AV to upgrade its functionality or for bug fixes in an embedded software installed in a data-generating sensor.
- *Service technician:* The service technician conducts maintenance services on autonomous vehicles and provides a report as a proof to detail its action on the vehicle.
- *Transport authority:* We assume that a transport authority would expand its role as a license-issuing authority and operators of roadside infrastructures such as stop signs and traffic lights by issuing smart roadside infrastructures and cryptographic credentials needed to secure vehicular communications. Secure vehicular communication allows an autonomous vehicle receive data from only vehicles with valid communication credentials.

We further classify the role of entities as proposers and validators to clarify which entities can send and validate data in the blockchain network. Proposers are entities that forward data in the blockchain or provide response to received requests. Proposers in our architecture include vehicle manufacturers, autonomous vehicles, and service technicians. Validators verify data communicated in the blockchain. Validators in our architecture include transport authorities and vehicle manufacturers. They are responsible for maintaining the blockchain and ensuring that identified malicious actions executed by interacting entities are broadcasted in the blockchain network. To achieve this, transport authorities utilize roadside units to ensure that communication occurs between legitimate vehicles in the vehicular communication network where vehicles and RSUs collaborate to exchange safety messages. Vehicle manufacturers request a vehicle to prove the status of its internal control prior to receiving a diagnosis update notification or after a scheduled maintenance has been conducted on the vehicle by a service technician. The internal control of the vehicle is the collective state of ECUs in the vehicle. It is represented by the Merkle tree root value obtained by computing the sum of hash of the firmware version running on in-vehicle ECUs as shown in Figure 9.2. Once the response is provided by the vehicle in the blockchain network, the validators compare the response value to the value in the blockchain. If the value differs, the vehicle is considered malicious and prevented from further vehicular communications to preserve the security of the vehicular communication network. In the next section, we discuss the data exchanges process between entities through transactions.

9.3.2.2 Transactions

The communication exchanges (i.e., transactions) reflect how a vehicle is added to the vehicular communication network, how the security of a vehicle's inter-

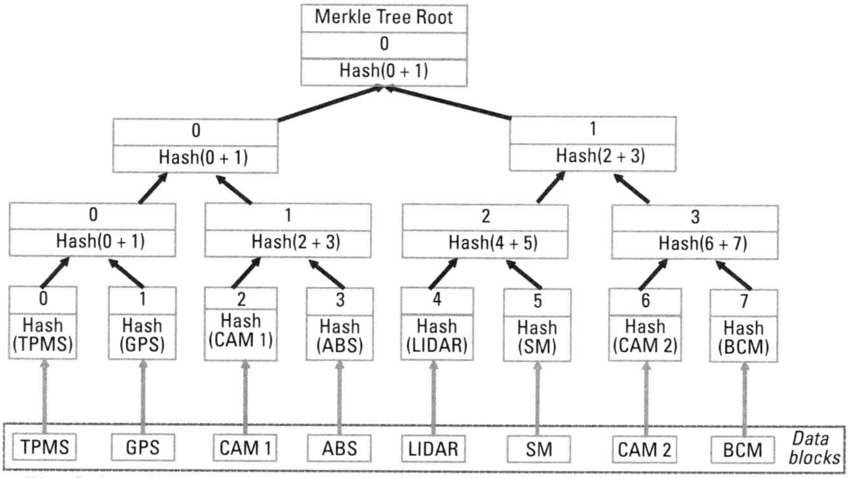

**Note: Each data block represent firmware running on an identified sensor.*
**hash(TPMS): is the hash value of the firmware on TPMS.*

Figure 9.2 Obtaining an ECU Merkle tree root.

nal control is assessed, and the actions executed by a service technician on a vehicle. To capture these actions, we define the following transactions:

- *Genesis transaction:* This follows the vehicle registration process. When a vehicle is initially assembled, the vehicle manufacturer computes the Merkle tree root value of ECUs in the vehicle and upon purchase, transmits this to the blockchain network. With this information, the transport authority creates a block for the vehicle and issues communication credentials to vehicle to enable secure vehicular communication. The block created by the transport authority for a vehicle consists of the block identifier (*BID*), the Merkle tree root of the vehicle (*MT*), the hash value of each ECU (*H(En)*), PK^+ of the vehicle (*PKv*), and PK^+ (PK_{TA}) and signature of the transport authority (TA) (δ_{TA}). The block identifier is used to keep track of changes to the Merkle tree value of ECU hash values. We present a detailed explanation about this in Section 9.3.3.
- *Proof of request and response transaction:* This transaction is created for two purposes: (1) to assess the security of a vehicle's internal controls, and (2) to ensure valid communications by restricting communications to legitimate entities. Consequently, we define two cases for the proof request and response: (1) the manufacturer of a vehicle assesses the integrity of a vehicle's internal controls; and (2) the RSUs confirm the legitimacy of a vehicle as it comes into its area of coverage. This transaction is a two-signature (multisig) transaction initially signed by a transaction initiator and completed by entity receiving the request. The structure of the multisig transaction is presented here:

$$[T_{ID}|\, Req\, |\, PK_{VM}/PK_R\, |\, \delta_{VM}\, |\, Resp\, |\, PK_V\, |\, \delta_V\, |\, T'_{ID}]$$

T_{ID} is the initial identifier of the request generated by the vehicle manufacturer. *Req* is the request to the vehicle to prove its internal state or to confirm its legitimacy. PK_{VM} is PK^+ of the vehicle manufacturer, while PK_R is PK^+ of the RSU. δ_{VM} is the signature of the vehicle manufacturer. *Resp* is the response provided by the vehicle, δ_V is the signature of the vehicle, while T'_{ID} is the transaction identifier computed by the vehicle when it provides *Resp* to *Req* and appends δ_V on the transaction.

In case 1, prior to sending an update instruction to a vehicle, the manufacturer sends a request to the vehicle to prove its internal state. The essence of this transaction is to identify if the internal control of vehicle has been compromised. The proof is for the vehicle to recompute its ECU Merkle tree root value. Once the response is received, the computed value is compared against the value in the blockchain. For case 2, when a vehicle approaches the area of coverage of an RSU, the RSU sends a proof of legitimacy request to the vehicle. Upon receipt of the request, the vehicle appends its signature and PK^+ and sends the response to the RSU. The RSU authenticates the vehicle by verifying the existence of PK^+ in the blockchain. If authentication fails, the vehicle is considered fake or malicious.

- *Maintenance transaction:* This transaction is initiated by a service technician when it executes a maintenance on the vehicle. The transaction consists of the maintenance details, the new firmware versions of ECUs if updated and the new ECU Merkle tree root of the vehicle. The transaction is broadcasted to the blockchain network and verified by validators before being stored in the blockchain.

  Transactions generated by trusted entities in our communication model are verified and validated by verifying the signatures of the entities using the verification component in the blockchain. However, transactions generated by untrusted entities such as the service technician and the vehicle are further verified. For a service technician, its signature is initially verified and then its action on the vehicle is verified by recomputing the ECU Merkle tree root value using the new hash values of updated/serviced ECUs. Next, the recomputed Merkle tree value is compared against the value computed by the service technician after maintenance. If the recomputed value is different from the value contained in the maintenance transaction, the malicious action of the service technician is identified. For a vehicle, when its response to the proof request is received in the blockchain network, verifiers initially verify if PK^+ of the vehicle exists in the blockchain. Next, to assess the internal state of the vehicle, the response from the vehicle is compared with the Merkle tree root value in the blockchain. If these values differ, the vehicle is considered malicious.

Having discussed the components of our security framework, in the next section, we present a compelling use case that describes how the security framework works.

9.3.3 Use Case

A smart vehicle is envisaged to randomly receive weekly over-the-air (OTA) update requests from a vehicle manufacturer [12]. Prior to receiving this request, the vehicle manufacturer needs to ascertain the integrity of the internal control of the vehicle. For this purpose, the vehicle manufacturer sends a *prove request* to the vehicle. This two-part signature transaction described in the previous section requires a vehicle to compute its ECU Merkle tree root value (MT). Figure 9.2 describes how MT is computed.

Once the vehicle receives the request, it verifies the signature of the vehicle manufacturer using its PK^+ (PK_{VM}). Upon completion, the vehicle executes the request by computing MT, including its PK^+ (PK_V), signing the transaction and computing a new transaction identifier (T'_{ID}). Next, it forwards the completed transaction to the blockchain for verification and validation. For verification, validators verify the signature of the vehicle and ascertain its legitimacy if its PK^+ exists in the blockchain. For validation, validators recompute (T_{ID}) and (T'_{ID}) to confirm the integrity of the multisig transaction when the *Req* was generated and when the *Resp* was provided. If this check passes, validators compare MT provided by the vehicle to the value in the blockchain. If the values are consistent, the vehicle is considered trustworthy and the vehicle manufacturer can send the OTA to the vehicle. The participating nodes must ascertain the legitimacy of the message generators, which protect against a Sybil attack where a malicious node floods the network with a large number of fake transactions. To achieve this (i.e., confirm the legitimacy of the vehicle while operational), RSUs send a request to the vehicle to prove its legitimacy when it comes to its area of coverage. The vehicle responds by producing MT and PK^+_V. If these values exist on the blockchain, the vehicle is considered valid.

To ensure that a service technician does not abuse its leveraged access to the vehicle, following successfully executed maintenance on ECUs, the service technician creates a maintenance transaction and broadcasts it to the blockchain. The maintenance transaction includes the hash of the ECU's update, the updated MT value, the signature, and PK^+ of the service technician and the transaction identifier. To verify the maintenance transaction, the signature of the service technician is verified and, if successful, transaction validation begins. The validation follows the steps taken by validators prior to sending an OTA update notification. The vehicle manufacturer sends a prove request to the vehicle, but, this time, the vehicle is requested to produce the hash values for its ECUs and the MT value. Upon receiving these values, validators initially map the hash values of ECUs updated by the service technician to the current value produced by the vehicle. If successful, validators compute the MT' value using the ECU hashes provided by the vehicle. Next, they compare the value obtained to the MT computed by the service technician. If the values are consistent, the vehicle's record in the blockchain is updated accordingly including its *BID* and its MT values.

9.4 Summary and Conclusion

In this chapter, we studied existing approaches towards securing smart vehicles, targeting vehicular network security, trust and reputation, privacy, and vehicular forensics. Furthermore, we utilized a risk analysis study to define priority properties necessary for securing smart vehicles. Based on this study, we discovered that the in-vehicle network presents a significant source of threat to smart vehicles and increases the opportunities of malicious adversaries to execute threats with critical impact to these vehicles and so threatens the security and safety of passengers. While existing works propose solutions that address security challenges of smart vehicles, they have not focused on the security of the in-vehicle network. Therefore, in this chapter, we present a blockchain-enabled countermeasure that achieves the requisite properties for securing smart vehicles. Furthermore, we present a compelling use case to highlight the efficacy of our proposal.

This chapter has explored blockchain's potential for smart vehicle applications, where typically the traceability is required on the state and context of the CPS entities themselves, which are highly dynamic. In the next chapter, we shift our attention to blockchain applications in CPS-supported supply chains. Supply chain applications involve the trading of discrete physical goods that are increasingly monitored by CPS sensors. In that respect, supply chains differ from the energy trading, where traded entities are pooled, and from smart vehicles, where the monitored entity is highly dynamic. We will see how the trade of discrete physical goods in supply chains presents its own unique challenges and opportunities for blockchain-supported CPS.

References

[1] Oham, C., R. Jurdak, and S. Jha, "Risk Analysis Study of Fully Autonomous Vehicle," https://arxiv.org/pdf/1905.10910.pdf.

[2] Greenberg, A., "Hackers Remotely Kill a Jeep on the Highway – With Me in It," Andy Greenberg Security, 2015, https://www.wired.com/2015/07/hackers-remotely-kill-jeep-highway/.

[3] Woo, S., H. J. Jo, and D. H. Lee, "A Practical Wireless Attack on the Connected Car and Security Protocol for In-Vehicle CAN," *IEEE Transactions on Intelligent Transportation Systems*, Vol. 16, No. 2, April 2015.

[4] Priyaasri, G., and M. Ranjith Kumar, "Vehicular Ad Hoc Networks Authentication Mechanism," *International Journal on Engineering Technology and Sciences*, Vol. 2, No. 3, March 2015, pp. 2349–3968.

[5] Oguma, H., A. Yoshioka, and M. Nishikawa, "New Attestation-Based Security Architecture for In-Vehicle Communication," *Proceedings of the IEEE GLOBECOM*, 2008

[6] Dorri, A., et al., "BlockChain: A Distributed Solution to Automotive Security and Privacy," *IEEE Communication Magazine*, 2017.

[7] Heijden, R., et al., "Blackchain: Scalability for Resource-Constrained Accountable Vehicle-to-X Communication," *SERIAL'17: ScalablE and Resilient InfrAstructures for Distributed Ledgers*, December 11–15, 2017.

[8] Li, L., et al., "CreditCoin: A Privacy-Preserving Blockchain Based Incentive Announcement Network for Communications of Smart Vehicles," *IEEE Transactions on Intelligent Transportation Systems*, 2018.

[9] Zhang, Y., et al., "OnionChain: Towards Balancing Privacy and Traceability of Blockchain-Based Applications," https://arxiv.org/pdf/1909.03367.pdf 2019.

[10] Malik, N., et al., "Blockchain Based Secured Identity Authentication and Expeditious Revocation Framework for Vehicular Networks," *2018 17th IEEE International Conference on Trust, Security and Privacy in Computing and Communications/12th IEEE International Conference on Big Data Science and Engineering*, 2018.

[11] Cebe, M., et al., "Block4Forensic: An Integrated Lightweight Blockchain Framework for Forensics Applications of Connected Vehicles," https://arxiv.org/pdf/1802.00561.pdf, 2018.

[12] Oham, C., et al., "B-FICA: Blockchain Based Framework for Auto-Insurance Claim and Adjudication," *Proceedings of the IEEE 2018 International Conference on BlockChain*, Halifax, 2018.

[13] Lei, A., et al., "Blockchain-Based Dynamic Key Management for Heterogeneous Intelligent Transportation Systems," *IEEE Internet of Things Journal*, 2017.

[14] Satoshi, N., "Bitcoin: A Peer-to-Peer Electronic Cash System," *Manubot,* 2019, https://git.dhimmel.com/bitcoin-whitepaper/.

[15] Madhusudan, S., and K. Shiho, "Intelligent Vehicle-Trust Point: Reward Based Intelligent Vehicle Communication Using Blockchain," arXiv Preprint, 2017.

[16] Yang, Z., et al., "A Blockchain-Based Reputation System for Data Credibility Assessment in Vehicular Networks," *2017 IEEE 28th Annual International Symposium on Personal, Indoor, and Mobile Radio Communications (PIMRC)*, 2017.

[17] Javaid, U., M. Naveed Aman, and B. Sikdar, "DrivMan: Driving Trust Management and Data Sharing in VANETs with Blockchain and Smart Contracts," *2019 IEEE 89th Vehicular Technology Conference (VTC2019-Spring)*, 2019.

[18] Shahid, A., et al., "Secure Trust-Based Blockchain Architecture to Prevent Attacks in VANET," *MDPI Sensors*, 2019.

10

Blockchain Applications in the Supply Chain

Sidra Malik
UNSW, Sydney, Australia

10.1 Introduction

In the previous two chapters, we covered how blockchain can support CPS applications for traceability of pooled physical assets (electrons) and highly dynamic and mobile CPS platforms (smart vehicles). In this chapter, we focus on the traceability of discrete physical assets within a network of untrusted participants. This scenario is embodied in supply chain applications where CPS sensors are used to monitor trade events and provide quality indicators on the traded assets. One of the key challenges in CPS-supported supply chain scenarios is trust in the trading entities and in the sensor data, as we discussed in Chapter 6, in order to ascertain the provenance and the quality of traded assets. Other important challenges relate to the commercial sensitivities of participants, which require privacy while sharing relevant data with others. Because of their global scale and the large number of participants, scalability is another key concern for supply chains. This chapter will explore these challenges and their respective solutions from the lens of supply chain applications. Before delving into the details, we provide an overview of supply chains and their features.

With the globalization of supply chains, there is a constant need to store the data for product provenance and traceability. The stored data can give information regarding the product history (i.e., origin, owners, manufacturing, and logistics). Counterfeit retail products, widespread epidemics, frauds, and organizational interoperability issues have further emphasized the need for reliable storage and efficient retrieval of information. Today, consumers show an unprecedented interest in how goods have been produced and handled before they make their way to shelves. Other supply chain stakeholders are interested in business outcomes that can benefit them in terms of higher sales and reputation. Despite realizing these requirements, the supply chain industry is still struggling to find the most suitable option to meet the expectations of both stakeholders and consumers.

Supply chains involve a complex web of relationships involving producers, manufacturers, retailers, and consumers. A simplified linear model of supply chain is shown in Figure 10.1. Each of these geographically dispersed stakeholders has isolated practices in terms of information storage and information sharing. Thus, collating data from disparate repositories becomes a significant challenge in supply chains, which makes traceability challenging. Some industries, such as agriculture or construction, are still mainly paper-based and underdigitized, which makes it hard to generate a full product story efficiently due to the organizational silos and missing information loopholes. However, farm-to-table traceability needs information such as ownership of goods, sales history, time and location details, and, critically, the information providers. Logging this information has the potential to address such consumer demands, to ensure safety, and to reduce fraud. However, it is critical to ensure that the log is not tampered with, which is a common malicious activity in the supply chain and may occur due to human error or an intention of name saving in the event of a product recall, for example, in sustainable supply chain products such as food and pharmaceuticals. Thus, traceability must ensure data integrity that is not just limited to the data content, but also includes the identity of the entity that logged the information at each stage of the product's handover, as we have seen in Chapter 6.

Global supply chains also face challenges regarding management, such as standardization of rules, automation of contracts, and auditing. In a supply chain, contracts may refer to any monetary, quality, time, and availability constraints involved in a trade. These contracts are mostly paper-based and failure to abide by contract terms often results in delays to determine accountable

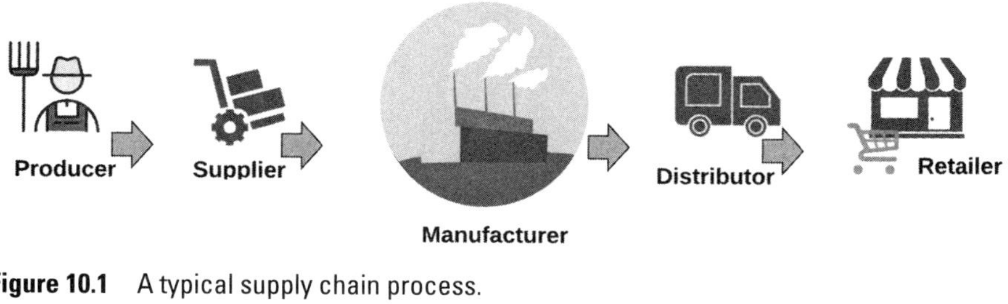

Figure 10.1 A typical supply chain process.

parties. In some cases, the legal disputes may take excessive time to get resolved. To facilitate both the material and information flow without delays, supply chains need automated governance.

The availability of reliable information has now become a requirement for consumers and stakeholders. This information, which increasingly relies on CPS sensors, may lead to valuable insights such as payment status for specific products, real-time traders' reputation, or identification of events and materials that may relate to the spread of foodborne diseases. Current databases that hold such information are siloed within their owning organizations, due to the inherent trade-off between information availability and the commercial sensitivity of releasing that information more broadly. In summary, current supply chain management (SCM) solutions do not provide sufficient provenance and traceability, automation, or information availability.

In recent years, blockchain has attracted tremendous attention in SCM to address the outlined challenges due to its salient features including immutability and security (see Chapter 3). The popularity of blockchain in SCM stems from its salient features, namely, immutability and transparency (see Chapter 3), given that the major challenges of SCM are integrity and traceability. The data of supply chain events such as origin, ownership history, and product condition can be stored in the blockchain in the form of transactions. The content of the transactions remain unchangeable due to the immutability of the blockchain. The transactions are permanently stored and publicly available to the blockchain users, which increase transparency.

Many industry-based solutions have started providing blockchain-enabled traceability solutions for SCM. Some solutions are focused on prompt commodity payments to growers (e.g., AgriDigital [1]), while others focus on the origin of raw material providers through smart contracts (e.g., Provenance [2]). In the shipping industry, ShipChain [3] blockchain can serve the purpose of information sharing while the goods are in transit. Researchers have also proposed several blockchain-based frameworks for SCM, which have increasingly incorporated traceability, IoT integration, and trust management.

In this chapter, we explore the potential benefits and challenges of applying blockchain in sensor-supported SCM. We study the existing approaches in literature and outline a roadmap to future trends. We first explore the blockchain requirements in SCM in Section 10.2, which describes the key requirements of blockchain technology tailored according to sensor-supported supply chains. Section 10.3 discusses the state of the art in blockchain-enabled SCM and categorizes the existing blockchain solutions across different domains of supply chain (i.e., traceability, scalability, platform selection, and trust). We study a traceable and trusted framework on a permissioned blockchain in Section 10.4. Finally, we outline the limitations of the existing systems and conclude the chapter by discussing research in Section 10.5.

10.2 Blockchain Requirements for SCM

In this section, we discuss the key requirements of blockchain-based systems for SCM. Applying blockchain in the SCM is not straightforward as it involves multiple application domains, each having different procedures in processing data and managing the supply chain process. The diverse regulations, globalization, and legal obligations across jurisdictions in a supply chain further complicate the design of a single global blockchain solution for supply chains. This highlights the demand for context-specific supply chain designs. Figure 10.2 shows the key requirements of SCM that highlights the demand for blockchain frameworks particularly optimized for SCM. The limitations can be categorized based on type (refers to the type of blockchain to be used, that is, permissioned, public, or private; see Chapter 3), role (refers to the realization of role-based data input and retrieval), application (refers to the stage in which the supply chain can be incorporated), and data (refers to the type of data logged in the ledger and the time when the data should be logged). We discuss each of these requirements below.

10.2.1 Blockchain Type

SCM solutions typically employ private or permissioned blockchains as they require significant control over the participants who can access the supply chain data. To ensure the transparency of the blockchain-based SCM and to ensure product provenance, certain information such as the origin of commodity is made publicly available. Information retrieval on a need-to-know basis is achieved using fine-grained permission management to third parties such as auditors.

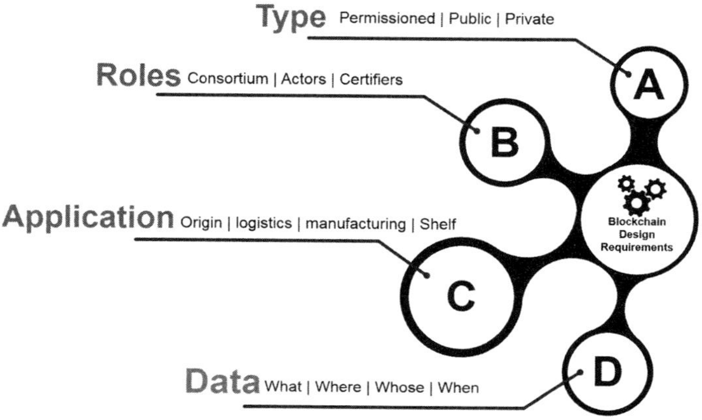

Figure 10.2 Blockchain design requirements in the supply chain.

10.2.2 Participants and Roles

A permissioned blockchain technology for SCM must be supervised and must define the relevant participant roles, besides actors, which include traders, suppliers, farmers, and retailers. Thus, before designing a blockchain solution, there is a need to define roles that can manage participation, access and grant rewards, or penalties for supply chain actors. A complete business model for permissioned blockchain design may include three types of roles:

- A consortium of organizers defines rules for access and blockchain technology vendors and allows participation. Unless the permissioned blockchain is proprietary to only one organization or private, the purpose of consortium is protection from dominance of the supply chain by a single party.
- Defining the role of supply chain actors adds granularity to access control rules and to the application of smart contracts.
- The supply chain industry thrives on certification and auditing from third-party certifiers such as certification authority, auditors, and registrars. These bodies are also either contributing to the supply chain data or need access to the traceability ands stakeholder's information on a frequent basis and hence are deemed important.

10.2.3 Application Stage

Blockchain can be applied at different stages of the supply chain (see Figure 10.1) that include:

- *Preproduction:* From the production of raw material to log original location and condition;
- *At-production and post-production:* In logistics to thwart counterfeit products in distribution and delivery;
- *Raw material to post-manufacturing:* After the raw materials have been used at a manufacturing unit to produce a final product that has a bar code;
- *Product story:* At the retailer's shelf for shelf-life assessment or to provide consumers with origin or manufacturing information.

10.2.4 Type of Data

Determining the type of data to be stored in the ledger is another significant design aspect. Supply chain data is increasingly sourced from CPS sensors. Supply chains are concerned about material and information flow, and blockchain can facilitate by providing a tamper-proof log along the following dimensions:

- The nature of product (what);
- The location (where);
- The quality of product (how);
- The quantity (how much);
- The ownership (whose);
- The payments (cost).

Blockchain transactions can be categorized as write or read transactions, which are employed by the users depending on the definition of data. Initially, the consortium decides on all the contractual obligations and smart contracts are written in the blockchain that contain the execution rules. Write transactions then automatically update the ledger when a supply chain event occurs. Read transaction facilitates in retrieving the information from the ledger.

This section studied the blockchain requirements for supply chain that involve deciding on the type of blockchain to use, the participants and their roles, the blockchain-relevant supply chain stages, and types of data to be stored on the blockchain. Next, we discuss the state-of-the-art solutions that leverage blockchain for supply chains which may or may not incorporate all these design considerations.

10.3 State of the Art in Blockchain-Based SCM

This section explores the value proposition of blockchain technology within manufacturing and distribution supply chains. We group blockchain-based SCM solutions based on their focus into four categories, namely: (1) blockchain platforms for SCM, which investigate the suitability of blockchain platforms for supply chains (see Section 10.3.1); (2) industry-led projects, which involve incremental adoption of blockchain into operations (see Section 10.3.2); (3) academic research projects, which propose more disruptive approaches for blockchain adoption in supply chains (see Section 10.3.3); and (4) trust-focused blockchain approaches, which quantify the degree of confidence in data being stored in blockchain-supported SCMs (see Section 10.3.4). We discuss each of these categories in further detail below.

10.3.1 Blockchain Platforms for SCM

Blockchain-based SCM frameworks are developed using various blockchain instantiations including Hyperledger (Fabric, Iroha, Sawtooth), Ethereum (both public and private), OpenChain, IBM Blockchain, MultiChain, Stellar, and EOS [4]. Besides using these platforms for proof of concept in supply chains, contributions are made in providing new consensus algorithms, integration of these platforms with off-chain databases and IoT devices, mechanisms to propose a new block in a product's ledger.

The authors in [5] introduced a new consensus algorithm optimized for supply chain requirements where each block contains the data of a particular product and is appended to the blockchain by the owner of the product. Each product post-manufacturing is identified using a unique near-field communication (NFC) tag along with a counter that is used for tracking the number of times that a tag is read across the chain. The manufacturer registers the product using a genesis block and a product's tag on the blockchain. Each time the product gets transferred, a new block is added to the product's chain and the receiver verifies the product details throughout its chain. The validator node then validates the new block using a global authentication algorithm that ensures the correct order of blocks in the product's blockchain, cross-checks the product's data in the previous and current block, and ensures that the source equals the destination in the previous block. The validation leader is a node mapped to each node proposing a new block and is responsible for random selection of validators for proposing a new block. The remaining validators wait on the final decision from the leader. The block commitment is based on a majority of two-thirds votes of the validators.

In [15], Sylim et al. focused on development of a distributed application (DApp) for a pharmaceutical surveillance system with a backend Distributed File System (DFS) and smart contracts. The general GS1 standard is adopted according to the U.S. Food and Drug Administration (FDA), which holds a role of data verifier. The supply chain actors include manufacturer, wholesaler, retailer, and consumer. As the data is distributed across DFS, smart contracts are populated in the blockchain that restrict access of data. DApps facilitate the data auditing process by comparing blockchain content with that of the DFS. If the cryptographic hashes do not match, DApp sends a notification to inform the participants of the discrepancy. The systems functionality can be tested using instances of simulated network on Ethereum or Hyperledger Fabric.

Having discussed the blockchain frameworks for supply chain, next we discuss industry-lead blockchain projects.

10.3.2 Blockchain as a Tool

Blockchain is widely applied by multiple companies in supply chain as a tool to log events during various application stages and to trace a product from initial stages of production to consumption. Most ventures employ blockchain to store the hash of data generated during multiple stages of supply chain as they argue that incorporating blockchain in SCM must not disrupt the traditional supply chain practices of data storage.

GS1 Australia's supply chain improvement project aims to increase the transparency of food raw material using GS1 standards coupled with blockchain for identification, data capture, and data sharing [8]. Their collaboration with The Commonwealth Scientific and Industrical Organisation (CSIRO) aims to incorporate GS1 EPCIS standards, integration of smart sensors, and packaging on blockchain. The idea is based on the notion that an implementation must be based on supply chain parties agreeing on a common way of uniquely identifying any item in terms of identity, location, and shipment. Since GS1 has been ensuring data definitions in supply chains for more than 40 years, the existing standards must be reused. Their aim is to share data across corporate boundaries while maintaining accuracy.

Provenance [2] is a pilot project that uses smart tagging technology along with blockchain and open data to track fish, wine, clothing, and other consumables. It helps businesses to add the origin and product journey information (i.e., providing the product story to their consumers). The verifications such as locally sourced, organic, charity support, and recycling are also added to the product. Their platform also allows the actors to share data in a permissioned environment.

Toyota [7] tracks its auto parts in various countries using blockchain technology. The applications and proof of concepts are being developed by the Toyota Research Institute (TRI) in three data areas: driving, car sharing, and user insurance. According to TRI, human driving data is important in development of reliable and safe autonomous vehicles. This data can be pooled in from vehicle owners, manufacturers, and fleet managers using blockchain [7]. TRI aims to create an awareness of blockchain technology among companies developing autonomous vehicle to expedite their motive of data acquisition through blockchain. We discussed the blockchain adaptation in smart vehicles in detail in Chapter 9.

Alibaba [8] introduced its blockchain traceability system called "Food Trust Network" in collaboration with Blackmores and Fonterra. Fonterra aims to ship its Anchor dairy products, while Blackmores plans to ship fish oil products. The goal of the pilot is promoting authentic products among consumers by allowing them a clear visibility over the chain. The authenticity information will be stored on product QR tags, which possess ownership, past and ongoing shipment details to ensure that the consumer trust in products is well placed. In December 2019, the group has also won two U.S. patents designed to make their consortium blockchain safe and fast [9].

In this section, we outlined a number of industry-funded projects to study the blockchain adaptation and applications in supply chain. Evidently, supply chain has received tremendous attention from various industries as a promising application of blockchain. We next study the blockchain frameworks employed for traceability of products.

10.3.3 Traceability Frameworks

Most of the existing blockchain-based solutions in SCM employ blockchain as a means to provide traceability and tracking, which are critical requirements in SCM. Blockchain-based traceability solutions trace various stages of the supply chain, including preproduction, manufacturer stage (production), logistics, and last-mile delivery to consumer. Some of these systems also target provenance where the value proposition of the product depends on its origin.

An online crowd validated shipment tracking framework is presented in [10], which leverages on state-of-the-art enterprise-based SCM systems. The framework consists of a set of private distributed subledgers and a single public ledger, as shown in Figure 10.3, which enhances privacy of the users. A private subledger is created for each shipment. The privacy-sensitive information such as the ownership of the goods, and the events are stored in the private blockchain and are only shared with the authorized participants. To ensure integrity, the hash of the private chain is stored in the public ledger. The monitoring events contain the geographical location of the shipment generated by the

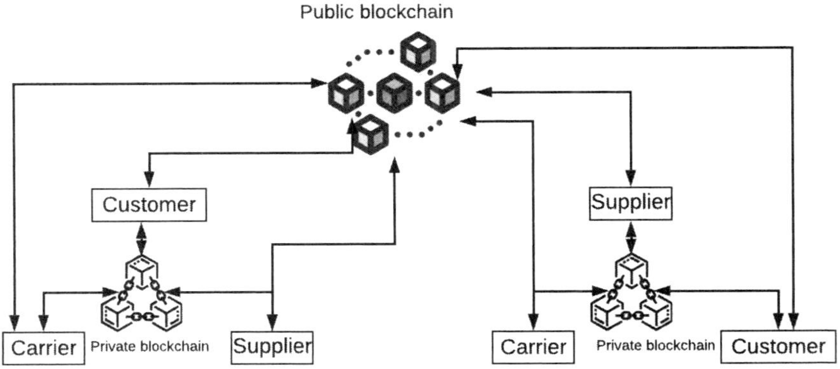

Figure 10.3 Event posting in public and private ledgers [10].

external monitors. These external monitors are triggered whenever a shipment is in close proximity and stores this information in the public ledger.

The authors in [11] presented an Ethereum-based blockchain framework to store information related to the manufactured goods, which potentially prevents counterfeiting. The product possession information is associated with an RFID tag to track the ownership of the product from manufacturers to their current owners. The system is based on logging and cross-checking electronic product codes (EPCs) of products and their manufacturer codes as they are being traded at each stage of the supply chain. To avoid counterfeit products, a seller must have a genuine EPC of product and hold ownership of the product. The latter is proved by knowing a valid manufacturer's address for the product that is conducted using the *getManufacturerAddress()* transaction. The *ShipProduct()* transaction can only be issued to the new buyer if the seller is the current owner of the product. To thwart the reusability of EPC and ownership information, a buyer may also generate a challenge message for the seller to sign and prove that current ownership. A product in a supply chain may be co-possessed by multiple owners. In such cases, a single EPC is registered against multiple owner addresses using the *enrollProduct()* transaction. To transfer a product with shared ownership, all the owners must agree as a prerequisite condition.

A blockchain–IoT-based Food Traceability System (BIFTS) is introduced in [12] that consist of three main phases. In Phase 1, an environment monitoring application is developed using multiple traceable resource units integrated with IoT sensors at container, batch, and item levels. The data collected from the IoT sensors is stored in a cloud storage. Phase 2 involves storing the log of the data in the blockchain in a traceable manner in real time. The consensus mechanism for stakeholders (considered as validators) uses shipment transit time, stakeholder assessment, and shipment volume. The blockchain information for a particular food item is then shifted to the cloud after the item has reached a final point of sales. Phase 3 involves providing the food quality information to a decision support system for shelf-life decay using the stored data and fuzzy logic. First, the data is collected from IoT sensors and provides information such as the shipment journey, milestones, and the environmental conditions of the product. This data is then preprocessed and structured in order to find the effects associated with temperature and humidity variation using mean kinetic temperature (MKT) and mean kinetic relative humidity (MKRH). MKT, MKRH, variation in total transit time, and scores from sensors are then used to evaluate dynamic food quality.

In this section, we discussed blockchain-based traceability frameworks. Although blockchain enables the participating nodes to trace data in the ledger, the authenticity of the stored data is not verified in blockchain. Next, we study the data reliability in blockchain.

10.3.4 Data Source Reliability

Blockchain provides trust among the untrusted participants using the consensus algorithm; thus, all participants agree on the state and data stored in the ledger. However, the authenticity of the stored data is not verified using blockchain (see Chapter 3). Supply chain data originates from multiple sources and validating the data across all sources becomes challenging, particularly because the data represents digital observations of physical events. This research direction is fairly new and is addressed in literature by incorporating reputation and trust mechanisms that have been extensively used in online e-commerce applications.

In [16], the authors introduced a novel model for the emission trading scheme (ETS) that incorporates blockchain technology with reputation-based trading to address fraud and management issues in the ETS. A reputation-based market segmentation and priority-value order mechanisms are designed that allow sellers with stong reputations to access more offers from the buyers and also to sort offers based on reputation and price. The reputations are calculated as a measure of emissions per product for an enterprise produce in the past. Participants with lower scores have a higher reputation. The second factor in reputation determination is the participant's strategy to achieve emission reduction. These strategies are based on the guidelines provided by the Clean Development program. Companies with more effective emission reduction strategies have better reputations. For example, companies A, B, and C use different emission reduction strategies. Company A uses cleaner fuel, while companies B and C adopt some additional measures along with cleaner fuel. Company B installed some new energy efficient equipment, while company C reconfigured some of their internal operations to improve energy efficiency. Then a third-party auditor, based on a company's strategy and final emission rates, assigns a reputation point score. For example, companies A, B, and C get a reputation point score of 2, 3, and 4. The companies with high reputation scores in the end will be provided and have an option to choose better options for trading.

Recently, IBM [13] developed cryptoanchors and tamper-proof digital fingerprints that can be embedded into products and linked to a blockchain as proof of a product's identity. These cryptoanchors are meant to be very small; for example, they can be found in edible magnetic ink shade that can be used to dye a malaria tablet. The IBM researchers endorse this notion of blockchain technology capabilities to infuse transparency, efficiency, and immutability but not guaranteeing the authenticity of physical goods. For instance, there is a need to reduce the fraud associated in critical supply chains such as life-saving drugs, which, in some countries, are 70% counterfeit. Similarly, 40% of the replaced automotive parts are not original. Within this context, cryptoanchors are claimed as a key solution. The inventors also argue that these cryptoanchors may not necessarily be embedded into a physical object. Some inherent features of products such as the DNA of rice corn or a weaving pattern in paper manufacturing or the molding of a specific metal when combined with artificial intelligence are unique enough to be considered as cryptoanchors.

In [14], a systematic discussion of incorporating trust is presented that addresses potential avenues to enhance trust and security in supply chains. Some examples include emphasizing the trust for SCM on a national or state level, at the trade association level, at the manufacturers' level, and at the consumer levels. The authors suggested that with the globalization of supply chains, SCM is becoming a huge challenge and achieving absolute trust in SCM is not a realistic expectation. However, it is achievable within a well-defined scope that one must define before aiming to provide a trust solution.

Overall, a blockchain-enabled supply chain management system must be capable of handling real-time supply chain data such as of trade events, IoT sensors, and third-party verifications in a way that is unforgeable, ensuring that these data events must be trustworthy and logging false data must be accounted for, providing availability of traceable information that does not violate the privacy of supply chain stakeholders, the governance is distributed such that no single organization is dominant on formulating the system rules, and the design and architecture should easily be compatible with the current blockchain platforms in terms of application, security, and scalability. In the following section, we introduce a blockchain-based approach for SCM that aims to address the outlined challenges.

10.4 Trustworthy Traceability in Supply Chains

Leveraging on the potential of blockchain as outlined earlier in this chapter, in this section, we present a solution that is holistic in terms of traceability, trust, and security and provides a sufficiently general application for any supply chain system. Our farm-to-fork traceability [15] and trust frameworks [16] are novel in that they cover basic design requirements defined in Section 10.2. We formulate our ideas around the use case of food supply chain as it has a potential to cover provenance and trust aspects in light of greater consumer demand of food security.

10.4.1 Blockchain Architecture

An overview of our architecture is shown in Figure 10.4. As outlined in Section 10.2, permissioned blockchains are mainly employed in the literature in supply chain use cases to support privacy and availability of information, the rightful contribution to the ledger, and making the information available for the right-

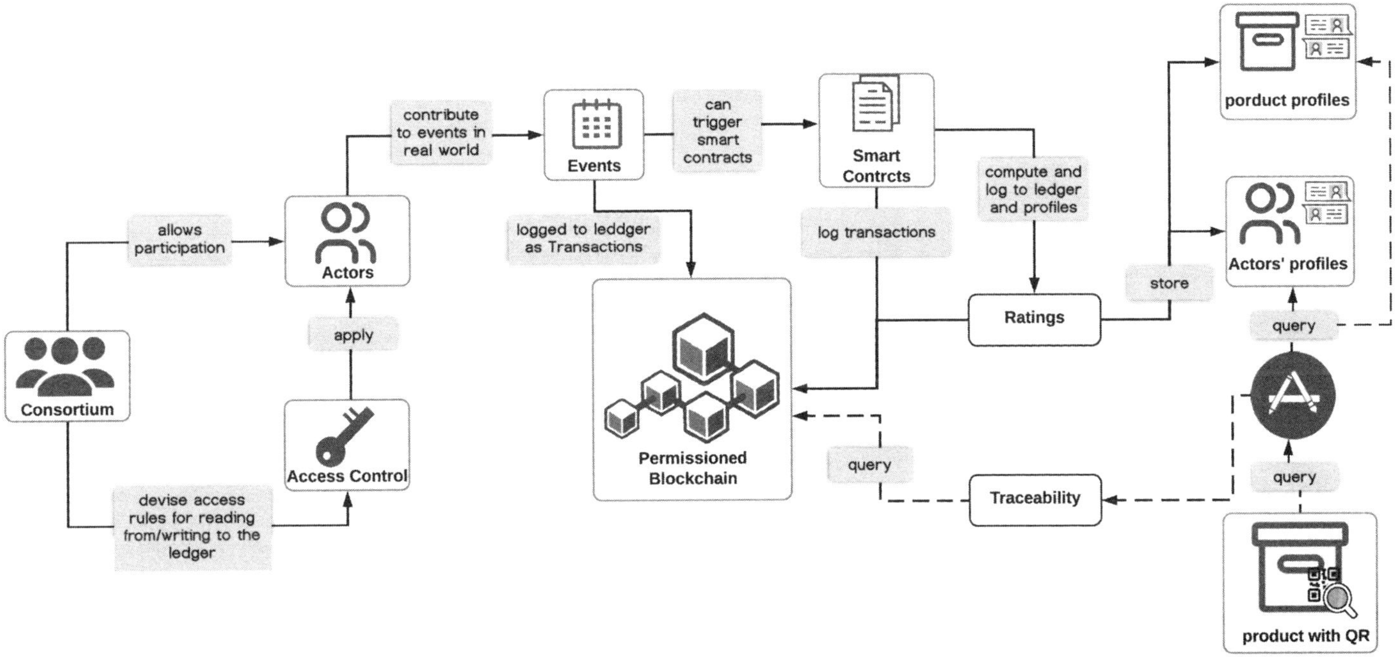

Figure 10.4 Architecture for trustworthy traceability.

ful participants. The consortium members agree on access rules to allow supply chain actors to participate in contributing data to the ledger. Once registered in the system, these actors are known by their digital profiles. The next step involves mapping the physical world of supply chain events with that of blockchain transactions. These transactions could be simple write transactions (log the event to ledger), which may trigger a smart contract, and query transactions (fetch data from blockchain). Having discussed the underlying permissioned blockchain, we next define two main functionalities of our system (i.e., traceability and trust). Our framework presents an end-to-end scalable approach to log farm-to-fork information on blockchain and thus achieve tracability. The trust module rates the actors based on the honest data contribution and the products based on their quality parameters defined in a smart contract. Finally, the query transactions via blockchain client applications can retrieve provenance, ratings, and product quality information. In the following sections, we elaborate on each module given in the architecture.

10.4.1.1 Governance

Our system is governed by a consortium based on government bodies to enforce the application and rules, for example, the Australian Competition and Consumer Commission (ACCC), third-party regulatory bodies such as the Food Regulatory Authority (FRA) for food safety and audit. It can also bring action on legislation and define penalties for participants if food standards are not met during the on-site physical inspections (e.g., refrigeration and packaging conditions for meat). The blockchain technology vendors meant to translate these rules into smart contracts deploy and manage the ledger (e.g., IBM, Amazon, and Skuchain) as a service platform. The inclusion of blockchain technology providers into the consortium ensures that the platform design and implementation is tailored to the business needs of the supply chain consortium. However, they are excluded from devising accessibility rules. The network is managed by the blockchain vendors' business/network administrators that connect the supply chain entities to blockchain. They define the business network model and choose a suitable platform. We use Hyperledger Fabric in our framework due to its key features including permissioned, ease of deployment, access control, and smart contract application. The supply chain entities store the event data to the ledger and maintain a static PK^+ in a business network after their registration from CA. The transactions issued by entities are verified by the blockchain platform host nodes such as validators in Hyperledger. These transactions are validated according to ACL and may trigger the relevant smart contracts in place with respect to a transaction. Note that ACL is subject to change only based on the two-thirds majority vote of the governing members. This may not be straightforward as said, but, due to the intertwined nature of supply chain, if the decision has to be made about an entity, it is then excluded from the decision board.

10.4.1.2 Overlay

We employ a sharded network to enhance the scalability of the blockchain where each regional zone is operated under a cluster of permissioned nodes known as validators. To reduce the overhead on the validators for verifying and storing new blocks, the validators in each cluster maintain a local ledger that stores the transactions of the nodes in the corresponding cluster. However, the farm-to-fork story may be dispersed across different local ledgers and the information is collated by keeping the ledger tag in each transaction while it gets logged. In Hyperledger, this can be fully implemented using the multiorganization Fabric Platform where channels can support the concept of shards. A set of common peers or validators connect all local ledgers to handle blockchain query requests using collating information from different ledgers as shown in Figure 10.5. More details and scalability evaluation of this model can be found in [15]. However, for this chapter, we will keep the discussion limited to one local ledger for ease of understanding the *Provenance* and *Trust* modules.

Having discussed the blockchain architecture, we next discuss the different types of transactions.

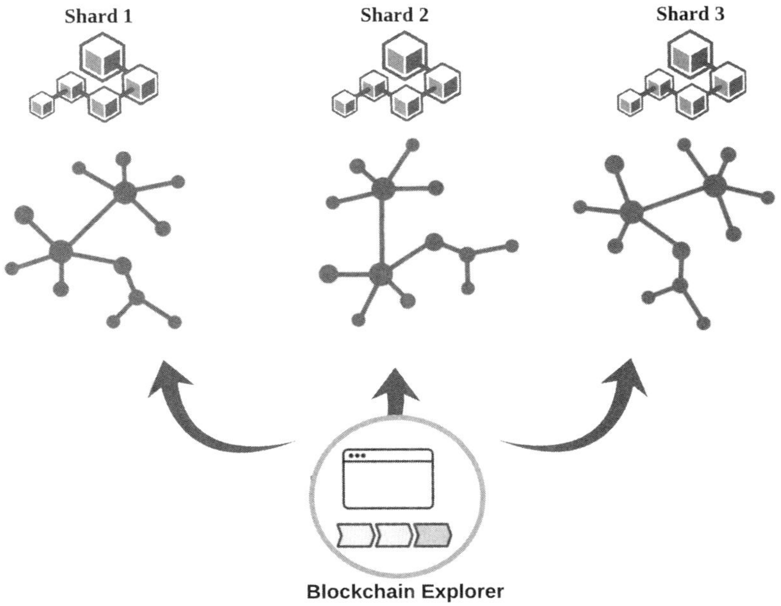

Figure 10.5 Sharded network architecture.

10.4.2 Transaction Vocabulary

Transaction vocabulary is the conversion of real-time supply chain data, which, unlike Bitcoin, is nonmonetary, into a blockchain transaction format. Data inputs in a typical food supply chain are considered from the IoT sensors, the traders in the supply chain, and third-party regulators. We simulate a commodity's journey from its production origin to its destination at the retailer's shelf using different transactions. We assume that every participating supply chain entity and new commodities have a digital profile (specific to Hyperledger and may vary with the use of different platforms), which is used to keep reputation and quality history, respectively. The transactions are described here:

- *New commodities:* Similar to genesis transaction, a new commodity is registered on blockchain using a create transaction, TX_{cr} by a primary producer:

$$TX_{cr} = \left[CID \,|\, H_{data} \,|\, ID_o \,|\, ID_{contract} \,|\, Sig_o \,|\, PU_o \right]$$

  where CID represents the commodity identity, H_{data} represents the hash of the commodity's data, which can be kept off the chain, ID_o represents ownership information, $ID_{contract}$ is the identity of a related smart contract if any, and, finally, the last two fields are the signature and public key of the primary producer.

- *Sensor data streams:* We assume that actors have installed IoT sensors to monitor food safety conditions such as temperature, location, and humidity. The information provided by these sensors is logged in the blockchain by a gateway on a periodic basis as the sensors themselves have low power capabilities. The structure of transaction used to log information, known as TX_{sens}, is as follows:

$$TX_{sens} = \left[CID \,|\, H_{data} \,|\, Sig_{device} \right]$$

  where CID is incorporated to the group TX_{sens} transactions related to one commodity, *data* refers to the sensor log, and Sig_{device} refers to the signature of the gateway node issuing the transactions. Whenever a TX_{sens} is logged, a smart contract is triggered that checks the temperature thresholds to which the commodity is bound and generates a commodity rating Rep_{sens}. Rep_{sens} specifies a rating assigned by the smart contract according to the degree of quality. This rating is stored on the commodity's profile (further discussed in Section 10.4.4).

- *Trading commodities:* These transactions are introduced specifically to log the change in ownership history. In other words, whenever a physical handover of commodity is executed between a seller and a buyer, a corresponding transaction known as TX_{tr} is stored on blockchain that is structured as:

$$TX_{tr} = \left[CID \,|\, H_{data} \,|ID_b | \,Sig_s \,|\, PU_s |\, Sig_b \,|PU_b \right]$$

where the *CID* and H_{data} is similar to that in TX_{cr}; however, the signatures of both the buyer and seller are included. In addition to this transaction, a rating, $Rep_{trader}(t)$, is attributed in seller's digital profile by the buyer [16]. Involving the buyer's assessment enforces the seller's honest behavior. However, in cases where a buyer may unethically give a wrong rating to a seller, a flag is generated against the buyer. The more flags a buyer has, the less trustworthy is the rating generated by the buyer.

- *Regulatory endorsements:* Endorsements are assumed to be generated by a regional authority responsible for executing food safety standard checks such as the FDA. These entities often conduct a site inspection of storage, production, cooling, or transportation (i.e., whether the sites follow a HACCP (Hazard Analysis and Critical Control Points) checklist). Based on third-party inspections, reports are generated to confirm the site safety standards. These reports are important and must be logged on the blockchain to protect their integrity in the events of product recalls or health safety checks. Hence, TX_{Reg} are introduced to log these third-party endorsements which are structured as follows:

$$TX_{Reg} = \left[ID_s \,|H_{data}|C_{type} \right]$$

where ID_s identifies the seller, as each site is owned by a particular seller. H_{data} is the hash of the inspection report, and C_{type} specifies the commodity type for which it was conducted. C_{type} is introduced for transaction distinction, as one seller may trade in more than one type of commodities. This transaction also generates a rating, $Rep_{reg}(t)$, for a seller that is stored on his or her profile. This rating is based on the regulator's assessment (i.e., the degree of food safety standards to which a seller's site adheres). Also, it is important to mention that these inspections are not frequent and are sometimes requested by the seller himself. In case the inspections are due in a specific time period, they must be requested by the seller to be conducted again. Aged regulatory endorsements will have a decreased weightage in the overall rating system until they are renewed next.

The generated ratings ($Rep_{sens}(t)$, $Rep_{trader}(t)$, and $Rep_{reg}(t)$) are used by the trust and reputation module (see Section 10.4.4) to calculate reputation and trust values for the supply chain entities and commodities.

- *Commodity's end of chain:* While TX_{tr} logs the commodity's trade along its supply chain, it is mandatory to log commodity's end of chain using a separate transaction TX_rec structured as:

$$TX_{rec} = \left[CID \,|\, Sig_r \,|\, PU_r \right]$$

where CID is again the commodity identifier and Sig_r and PU_r are the signature and the public key of the receiving retailer. This transaction is important for the following reasons:

  - *Security:* This transaction will help in identifying fake commodities registered on blockchain as it is unlikely that a nonexistent commodity will have a progressive chain and thus could be easily distinguished if the chain is neverending.
  - *Commodity rating:* A commodity upon reaching at its destination will be evaluated for its quality (e.g., if the temperature conditions were met throughout its journey). This will generate an overall rating using a quality smart contract (see Section 10.4.4) for commodity, updated in the commodity's digital profile that can be made available to the final buyer via blockchain application.

In the past two sections, we outlined the details of the fundamentals of our framework and different types of transactions. Next, we describe how the framework is used to provide traceability in supply chains.

10.4.3 Traceability Module

Recall from Section 10.3.1 that in our architecture traceability refers to the availability of information for a commodity's farm-to-fork chain. The aim of our framework is to minimize the delay experienced by a user, which is challenging as many commodities are combined at an intermediate level such as a manufacturer to produce a packaged product. For example, a consumer of a fresh product such as grapes or strawberries could validate the originating farm in minimal time and evaluate the history of storage conditions logged in the blockchain.

Thus, reconciling information after the commodities of various origins have been mixed, referred to as the manufacturer's stage in the rest of this chapter, is a great challenge. The prepackaged food is often labeled in the form of batch identifiers where each batch ID depicts the IoT from which it is sourced. Post-packaging, these items with different identifiers are combined and labeled again with EPCs or bar codes, which include the manufacturing and expiry dates. Thus, on blockchain, a product is often identified with a bar code that is only assigned by the manufacturer. To keep the farm-to-fork story intact on blockchain, the batch IDs associated to a commodity or the individually stored items must be connected to the final packed product labeled with bar codes. Recall that on blockchain we only have a record of individual commodity (TX_{cr}) comprising the *CID* or batch ID (data part of the TX_{cr}). In order to link the *CID*s with the final packaged code, we introduce produce transaction. This transaction binds the individual batch IDs with those of the final packaged product. The transaction process is further illustrated in Figure 10.6.

The batch IDs depicting different lots of commodities are constituted in ledger by individual TX_{cr}. The final product may contain different lots of same commodity A such as (a1, a2, a3) combined with other commodities such as B and C. When the different lot items are combined at any stage, the *produce* transaction must be generated, which is given by:

$$TX_{pr} = \left[ID_p \middle| ID_i \middle| H_{data} \middle| Sig_s \middle| PU_s \right]$$

ID_P is the product identifier, that is, the bar code and the ingredient identifier, ID_i, identifies the batch IDs of different commodities. Data may contain basic information regarding batch, weight, location, sensor, or other information. A valid provenance information for any raw material only exists when ID_P and ID_i correspond to a valid TX_{cr}.

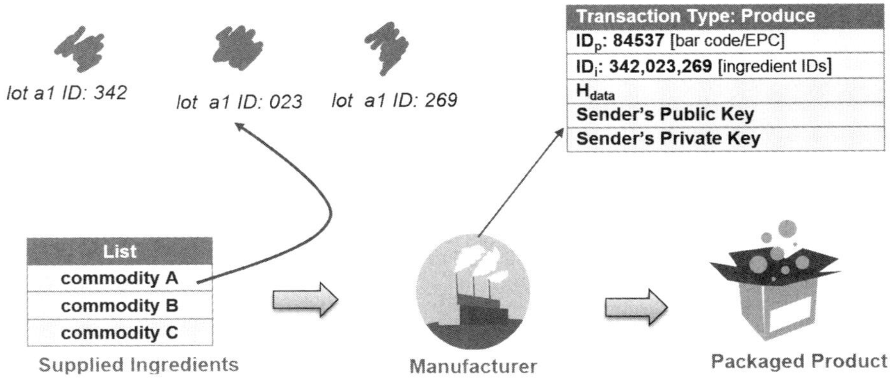

Figure 10.6 Produce transaction.

Produce transaction may also be linked to TX_{sens} or TX_{reg} using ID_p and seller's digital identity, respectively. Having discussed the traceability framework, we next discuss trust management in our framework.

10.4.4 Trust Management Module

As outlined earlier in Chapter 6, blockchain provides data integrity. However, the correctness of the information logged on the ledger is not guaranteed; thus, the authenticity of the stored data in blockchain cannot always be assumed to be trusted. Recall from Section 10.3.4 that trust is a critical requirement as SCM events are physical and the correctness of the data cannot be verified by digital means (see also Chapter 6). Incentivizing the supply chain participants to contribute honestly and accounting for logging false information can improve the trustworthiness of the system.

We propose a reputation model to be embedded as an extra layer on blockchain to provide trust. This layer is executed by two smart contracts: *Reputation Contract*, which manages the reputation of sellers, and *Quality Contract*, which manages the reputation of the commodity based on its quality. In the following sections, we explain these smart contracts and how the trust layer works in coordination with these contracts to enhance data authenticity.

10.4.4.1 Smart Contracts

Recall from Section 10.4.2 that transactions are associated with ratings that are generated for supply chain entities and commodities in a secure and automated way using smart contracts. The ratings are used to calculate the overall trust values for the supply chain entities, which together with the application layer sought rewards and penalties. Since our implementation [16] is based on Hyperledger, we explain the working of these contracts accordingly.

Smart contracts are installed by business network administrators that may belong to a single organization or a group of organizations. Moreover, the local access rights to install these contracts can be defined by the message service provider (specific to Hyperledger).

- *Quality contract:* There could be multiple quality contracts installed depending upon the quality conditions that a commodity requires; for example, the transportation of meat may require different temperature thresholds and warning conditions, whereas fresh produce such as grapes could have different quality requirements for transit. It is compulsory for each commodity to adhere to a quality contract while registration uses TX_{cr}. Once the commodity is registered, it is being monitored against the contract conditions regularly whether it is in storage or in transit. The contract enlists the following criteria:

- *Quality rating criteria:* That is, temperature thresholds for the commodity that include:
  - The maximum upper and lower bounds of the required temperature in which a commodity is considered safe;
  - The temperature damage threshold, which when exceeded, indicates that the commodity is no longer in a consumable state.
- The contract identifier, $ID_{contract}$.

This smart contract is triggered every time a temperature reading is logged using TX_{sens} for the commodity. Based on the threshold conditions in the contract's description, two types of outputs are generated in response to TX_{sens}: (1) warnings if the commodity's temperature damage threshold has been reached, and (2) ratings of the commodity based on the temperature reading. Ratings are further elaborated as the instantaneous ratings $Rep_{sens}(t)$ generated whenever a commodity is traded using TX_{tr}. These ratings are stored with every trade event until the commodity reaches the final retailer. The quality contract generates $Rep_{sens}(t)$ based on the temperature history of the commodity stored as $Rep_{sens}(t)$ over its transit journey.

For n number of trade transactions for a commodity, the respective ratings are stored in the commodity's profile. Recall from Section 10.4.2 that once the commodity reaches the final destination, the retailer generates a receipt of commodity's end of journey using TX_{rec}. This transaction invokes the quality contract to generate the final rating of the commodity, Rep_{sens}

$$Rep_{sens} = \left[Rep_{sens}\left(t_0\right),\ Rep_{sens}\left(t_1\right), \ldots\ldots,\ Rep_{sens}\left(t_n - 1\right) \right]$$

The final commodity rating can be made available to the consumer of commodity using a QR code that represents the data in the blockchain.

- *Rating contract:* After each trade, TX_{tr} is invoked and updates the rating of the seller. As outlined in Section 10.4.2, each trade event contains the respective seller's rating from the buyer ($Rep_{trader}(t)$), the regulatory endorsements ($Rep_{reg}(t)$) for the seller, and the reputation score of the traded commodity ($Rep_{sens}(t)$). The rating contract at this point generates an overall seller's rating as $Rep_{seller}(t)$, which is calculated as a weighted sum:

$$Rep_{seller} = \left[w_1 \times Rep_{sens}\left(t\right),\ + w_2 \times Rep_{trader}\left(t\right) + w_3 \times\ Rep_{reg}\left(t\right) \right] \quad (10.1)$$

where w_1, w_2, and w_3 are the weighting factors for each reputation contributor. The quantification of these weights is application-specific and dependent on the rules defined by the consortium. For example, in the case of less frequent sensor data, w1 can be reduced, or, if regulatory endorsements are outdated, $w3$ can be reduced. $Rep_{seller}(t)$ is then stored in a seller's profile for a particular trade event. These scores are then aggregated to give an overall rating and trust value detailed in the following section.

- *Reputation and trust management:* The rating contract facilitates the management of a seller's rating for a commodity over a certain time period. To calculate the overall seller rating, a reputation model can be chosen based on any aggregation function such as the mean, median, or beta-reputation model. For our trust management module, we chose a time-varying and amnestic reputation and trust model [16]. By time-varying adaption, we mean that the ratings pertaining to recent supply chain events are given more weightage than the ones in older times.

The overall trust score of a seller involves two steps: (1) overall reputation score $R(t_n)$ of the trader based on the historic $Rep_{seller}(t)$, and (2) calculating the overall trust score based on $R(t_n)$ and other features specific to the supply chain application. The two steps are described in detail here:

- *Step 1:* $R(t_n)$ for a trader in the previous times $Rep_{seller}(t_0)$, $Rep_{seller}(t_1)$, ..., $Rep_{seller}(t_n)$ is given as:

$$R(t_n) = \sum_{t=t_0}^{t=t_n} Rep_{seller}(t_0) \times \beta(t_n - t) \tag{10.2}$$

  where $Rep_{seller}(t_0)$ refers to the initial trade event of the trader and $Rep_{seller}(t)$ corresponds to supply chain event happening at time t. $\beta(t_n{-}t)$ is the forgetting factor, given by any decaying function (e.g., $\beta(t) = e^{-f(t)}$), which can enable the reputation score to evolve with time. This emphasizes the recent events in supply chain to have more affect in the overall reputation, $R(t_n)$ of the system. This score is limited to one type of commodity. If a trader is in business with multiple commodities, the reputation is separately calculated for each commodity and stored in distinct digital profiles.

- *Step 2:* This step involves calculating the trust of the seller based on the information logged on the ledger. Note that trust and reputation are not interchangeable in our definition of trust. We consider reputation as one of the major contributors to the trust of the system but may include other factors as well.

Upon joining the network, a new trader has no past reputation or trust values and hence is assigned the initial trust score $Trader(t_0)$ to be $Trust_{min}$. $Trust_{min}$ is the mininimum trust score that each entity maintains to participate in the system. The trust score evolves as the entities trade in the system and each entity should maintain $Trust_{min}$ to be allowed to participate in the Trust-Chain network. The overall trader's trust value $Trader(t_n)$ is based on the overall reputation $R(t_n)$ and other contributing features denoted by f_1, f_2, ..., f_N, (for example, number of sales). The equation to represent this is:

$$Trader(t_n) = \alpha_0.R(t_n) + \alpha_1.f_1 + \alpha_2.f_2 + \ldots\ldots + \alpha_N.f_N \quad (10.3)$$

where α_0, α_1, ..., and α_N are the weighting factors for computing trust and are determined by the governing board and the business network administrator. To further illustrate the use of above-mentioned features, let us consider the trader's number of successful transactions as a feature example. The values for feature scores are given in Table 10.1, which again is decided by the consortium.

10.4.5 Application Product Queries, Penalties, and Rewards

The traceability and trust information logged on the ledger can be retrieved using "query" transactions. These queries are invoked by the application layer to attain either provenance or quality of commodities or trust information of the traders. Based on the query results, the following features can be ascertained using our modules:

- *Provenance:* In the traceability module, with batch IDs of raw materials being tied to the final product, the origin of the packaged item can be determined.

Table 10.1
Example of Feature Scores

No. of Successful Transactions	Feature Score (f_1)
0	−1
1 to 3	0.5
4 to 6	1.5
>6	2

- *Rewards:* Supply chain entities, having a higher trust value, are termed honest and must be incentivized to maintain their honest behavior. Queries can be formulated to sort, choose, and publish the traders based on their reputations and trust score.
- *Penalties:* The traders with low reputations can be revoked from participating into the network for a certain period of time. The revocation list itself can incentivize the other traders to participate honestly in the system.

10.4.6 Case Study

In this section, we provide a case study of the proposed architecture based on the milk and cocoa used in chocolate bar. Figure 10.7 depicts multiple steps involved in creating a product ledger. A product ledger is the chain of transactions involved in the life cycle of a product that includes trade transactions TXtr, produce transactions TXpr specific to a single manufacturer, and corresponding TXcr with respect to each key ingredient in TXpr. As an example, milk in Figure 10.7 has a product ledger of size 6. In this case study, we assume that the blockchain size is 10 transactions. In our framework, the information about each product is stored in a shared ledger (see Chapter 3). To enable the users to trace the history of transactions, a global validator (BC_{global}) is defined. The users can query BC_{global} to receive historical information of a product.

To retrieve the information about a product, the user requires a T_{ID} and the product identifier, which ideally should be included in a QR code that is displayed on the product. Consider the example shown in Figure 10.7. The consumer scans the QR code that queries BC_{global} to locate historical information about the product. BC_{global} first collects T_{ID} of the TXtr corresponding to the trade between the distributer and the retailer (step 1). $P.T_{ID}$, the ID of the previous transaction, of the transaction in step 1 leads to the TXtr trade between the manufacturer and the distributer (step 2). TXtr in step 2 connects to TXpr, which is generated by the manufacturer (step 3). TXpr is linked to two TXtr, which technically are the transactions corresponding to the trades between the manufacturer and the suppliers of the raw materials (step 4). Step 5 includes the trade between the supplier and the manufacturer of milk and cocoa. Following these steps, the product ingredients can be traced back to the farmers. BC_{global} returns information such as time, date, and location information associated with the outlined transactions to the consumer (step 7).

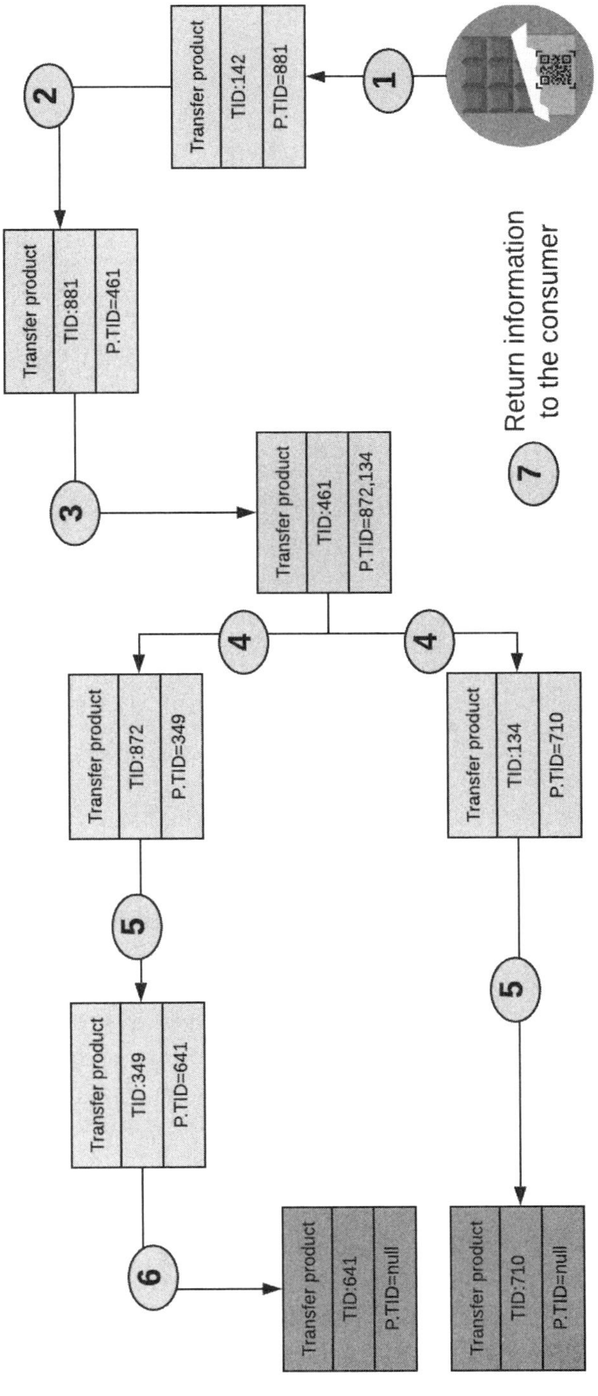

Figure 10.7 Provenance information for cocoa.

10.5 Critical Analysis

In this section, we provide a critical evaluation on the existing works studied in this chapter (see Sections 10.3 and 10.4) to highlight their strengths and weaknesses in meeting supply chain requirements. The results of evaluation are summarized in Table 10.2, which shows that while existing approaches address some of the SCM management, our ProductChain and TrustChain approaches are more comprehensive in tackling these challenges.

10.6 Summary and Conclusions

In summary, the existing blockchain-based trust and traceability systems do not cover all the requirements of supply chains. The formulation of design requirements, defining the scope within the application use case, and scalability are important considerations. We designed ProductChain [15] and TrustChain [16] considering the supply chain-specific challenges as outlined above and take into account the multiple data observations, automation of rating calculations, accountability mechanisms, a quality security analysis, and performance evaluation in terms of network throughput and latency. There is still room for improvement to provide more comprehensive traceability and trust in supply chains with blockchain. We may conclude that while blockchain may not be the only solution for the supply chain industry, its incorporation in conjunction with traditional databases and IoT can address many existing challenges of SCM.

In Chapters 8 and 9 and this chapter, we covered blockchain in the context of CPS application that trace or monitor physical assets, ranging from pooled physical assets, to mobile CPS platforms and discrete physical assets. We saw that these applications involve challenges that typically arise from the flow information across the interface between physical and digital domains. In the next chapter, we shift gears to the use of blockchain for supporting trading of digital assets, namely, IoT data in a marketplace context. While the focus on digital assets relaxes some of the challenges that we observed relating to physical assets, it involves other challenges relating to decentralization, privacy, and real-time operation of CPS sensors, which involves challenges relating to the persistence of CPS devices.

Table 10.2
Comparison of Blockchain-Based SCM Approaches

● Highly focused	◐ Partially focused	○ Low focused	✓ Incorporates	✗ Does not incorporate

References

[1] "AgriDigital Australia," https://www.agridigital.io/products/blockchain, accessed April 2020.

[2] "Provenance: Every Product Has a Story," https://www.provenance.org/, accessed April 2020.

[3] "ShipChain," https://shipchain.io/, accessed April 2020.

[4] "Top Blockchain Platforms to Watch Out in 2019," https://hackernoon.com/top-blockchain-platforms-to-watch-out-in-2019-aa80e336a426, accessed April 2020.

[5] Buchman, E., "Tendermint: Byzantine Fault Tolerance in the Age of Blockchains," (Doctoral dissertation), University of Guelph, https://atrium.lib.uoguelph.ca/xmlui/handle/10214/9769.

[6] Sylim, P., et al., "Blockchain Technology for Detecting Falsified and Substandard Drugs in Distribution: Pharmaceutical Supply Chain Intervention," *JMIR Research Protocols*, Vol. 7, No. 9, 2018, p. e10163.

[7] "Toyota Looks to Blockchain for Secure Communications," https://www.engineersaustralia.org.au/News/toyota-looks-blockchain-secure-communications, accessed April 2020.

[8] "Alibaba Pilots Blockchain Trial for Supply Chain Transparency with Australian Products," https://www.smartcompany.com.au/startupsmart/startupsmart-technology/alibaba-pilots-blockchain-trial-for-supply-chain-transparency-with-australian-products/, accessed April 2020.

[9] "Alibaba Patents Would Secure, Accelerate Its Consortium Blockchain," https://www.coindesk.com/alibaba-patents-would-secure-accelerate-its-consortium-blockchain, accessed online April, 2020.

[10] Wu, H., et al., "A Distributed Ledger for Supply Chain Physical Distribution Visibility," *Information*, Vol. 8.4, 2017, p. 137.

[11] Toyoda, K., et al., "A Novel Blockchain-Based Product Ownership Management System (POMS) for Anti-Counterfeits in the Post Supply Chain," *IEEE Access*, Vol. 5, 2017, pp. 17465–17477.

[12] Tsang, Y. P., et al., "Blockchain-Driven IoT for Food Traceability with an Integrated Consensus Mechanism," *IEEE Access*, Vol. 7, 2019, pp. 129000–129017.

[13] "Combating Fraud with Blockchain and Cryptoanchors," 2019, https://www.research.ibm.com/5-in-5/crypto-anchors-and-blockchain/.

[14] Kshetri, N., "1 Blockchain's Roles in Meeting Key Supply Chain Management Objectives," *International Journal of Information Management*, Vol. 39, 2018, pp. 80–89.

[15] Malik, S., S. S. Kanhere, and R. Jurdak, "ProductChain: Scalable Blockchain Framework to Support Provenance in Supply Chains," *2018 IEEE 17th International Symposium on Network Computing and Applications (NCA)*, Cambridge, MA, 2018, pp. 1–10.

[16] Malik, S., et al. "TrustChain: Trust Management in Blockchain and IoT Supported Supply Chains." *2019 IEEE International Conference on Blockchain (Blockchain)*, 2019, pp. 184–193.

11

Blockchain Applications in IoT Data Marketplace

Pooja Gupta
UNSW, Sydney, Australia

11.1 Introduction

This chapter focuses on the exchange of digital assets from CPS, namely IoT data. IoT encompasses a wide range of sensors and devices that capture sensor observations from the environment that are shared with service providers (SPs). In conventional IoT systems, the data is stored by the SP of a device leading to the creation of data silos where data access is only restricted to the SP and the inability to share it with other parties. Given the significant value of the data produced by the IoT devices, the notion of an IoT data marketplace has attracted tremendous attention as it enables the IoT users to monetize, trade, and share the historical or real-time data generated by their IoT devices with other participants in the IoT ecosystem. The data marketplace addresses the issue of data silos and incentivizes the users to share their data with SPs or other parties that require access to the data. The fundamental components of a data marketplace [1] are as follows:

1. *User registration and authorization:* This allows users to register themselves in the marketplace either as sellers or buyers. A seller owns data while a buyer is interested in purchasing a particular type of data. The seller may implement different granularities of access and only allow

certain buyers to access a data stream for a desired duration or at a required interval.

2. *Data discovery and selection:* This allows buyers to search and select sellers' devices based on their desired data type (e.g., traffic data or solar panel data).
3. *Data price model:* This allows the seller to select a pricing strategy with typical options being fixed, tiered, dynamic, or negotiated pricing. In a fixed pricing model, the price is fixed for a particular duration, while in tiered pricing, the price is fixed but includes multiple categories. Dynamic pricing changes the price over time, while in negotiated pricing, the price is actively negotiated at the time of sale.
4. *Contract management:* This is employed by the seller and buyer to manage the trade agreement. A contract usually consists of terms and conditions that govern the quality of data and the associated costs.
5. *Rating mechanism:* This function is employed by the buyers and sellers to rate each other, which establishes a level of trust.
6. *Data metering:* This is used to measure particular parameters (e.g., amount, frequency, and type) associated with an ongoing trade for billing purposes.
7. *Billing and payments:* This manages the payment process for trades. The marketplace may incorporate different mechanisms of payment (e.g., advance payment, pay-as-you-go, payment at predetermined intervals).
8. *Real-time middleware:* This incorporates different functions related to real-time data streaming such as data routing, resource management, and encryption. When such functions run on CPS nodes, they can raise the issue of persistence that we presented in Chapters 1 and 2.

Conventional data marketplaces rely on centralized platforms where a central TTP oversees the trade to ensure that both parties commit to their obligations, which results in several limitations. The first limitation was limited scalability, as the TTP conducts all management operations and controls every aspect of a trade from offer listings to price discovery, product search, data storage, transacting the trade, data dissemination, and buyer/seller feedback. With billions of connected IoT devices, centralized governance can be a bottleneck. Other limitations were limited privacy as the TTP has full visibility on the trade history of sellers and buyers and the associated data, central point of failure, and low security as the TTP may be compromised by hackers, which leads to data leakage.

To address the outlined challenges, blockchain has attracted significant attention to serve as a distributed framework for IoT data marketplace due to its salient features including transparency, decentralization, security, anonymity, and immutability (see Chapter 3). A node in a blockchain network is known by a pseudonym, which is its public key, instead of the real-world identity, hence enhancing user anonymity. Every time a user transacts IoT data, the corresponding transaction is recorded in the blockchain that creates an auditable log of the transactions, which supports trust in every transaction. The existence of an encrypted, perpetual, and validated record of transactions via blockchain increases confidence in buyers and sellers to exchange their sensitive IoT data. The multidimensional benefits of the blockchain in IoT data marketplace are summarized in Figure 11.1.

In this chapter, we first study existing blockchain solutions to facilitate marketplace functions in Section 11.2. In Section 11.3, we propose a blockchain-based IoT data marketplace framework, which specifically addresses the

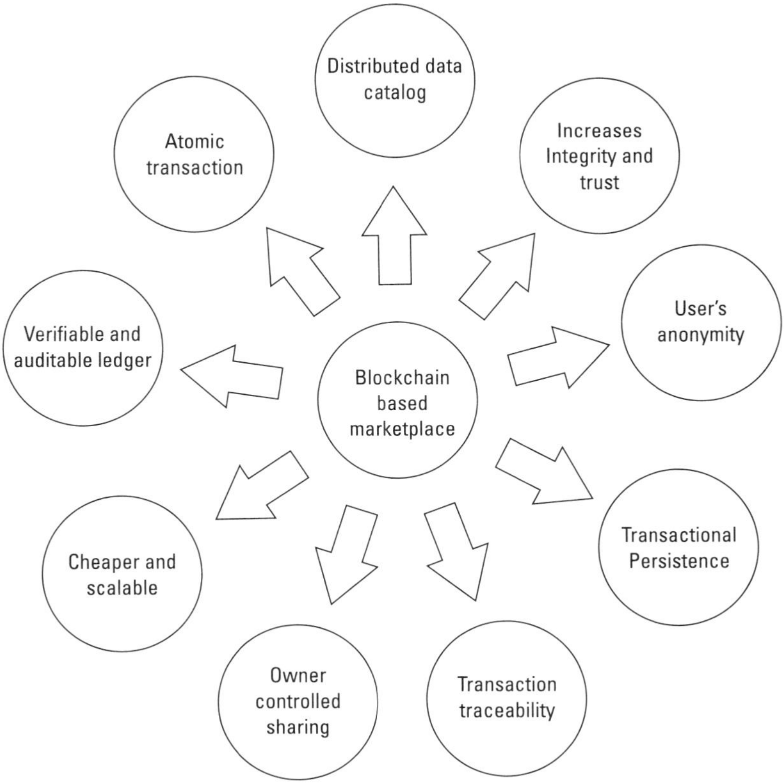

Figure 11.1 The fundamental advantages of the blockchain in data marketplace.

issues with the state-of-the-art approaches. Section 11.4 provides a critical analysis of the existing solutions, and finally Section 11.5 concludes the chapter.

11.2 Towards Blockchain-Based Data Marketplaces

In this section, we study the existing blockchain-based approaches for realizing the IoT data marketplace. We categorize such solutions based on how they use blockchain for realizing different marketplace functionalities in five categories: trade transaction management, distributed data catalog, hybrid centralized-decentralized architectures, decentralized data storage and access mechanisms, and agreement instantiations. We highlight the core features of each approach and discuss some example frameworks.

11.2.1 Trade Transaction Management

The purpose of a marketplace is to facilitate trade of a particular item between users. A trade transaction is defined as an exchange of value between a buyer and seller that involves the user's communication, negotiation, execution, settlement, and payment. In a blockchain-based trade management framework, the entire trade transaction history is recorded in the blockchain, which enables the participants to trace the history of the transactions based on the transaction time stamp and have full visibility of the communications and data exchanges, leading to a high level of transparency. Benefitting from the inherent immutability offered by blockchain, when a transaction is confirmed, it is impossible for either party (i.e., seller or buyer) to modify the content or data, which provides a trusted trading platform.

The authors in [2] introduced a secure publish/subscribe (SPS) service to protect the anonymity of the users and confidentiality and achieve fair payment in exchanging CPS data. The SPS protocol considers the behavior of a participant in deciding on the reliability of data generated by that participant. SPS defines a reputation factor R for each participant where a participant is believable if its corresponding R is greater than a threshold; otherwise, it is not fully trusted. Each user employs a public key (PK^{+}) as its identity that introduces a level of anonymity. As shown in Figure 11.2, SPS consists of multiple publishers, sensors, and subscribers. It involves the following steps, which are implemented using Solidity on Ethereum. The first step is that the publisher collects the data from the device and publishes a topic describing the captured data type to Bitcoin, which is used as the underlying blockchain. Untrusted publishers are required to deposit a predetermined value before publishing their topic, which protects against malicious publishers that may inject fake information. The second step is that a subscriber leverages the ElGamal encryption scheme to encrypt the topic and sends a subscription request to the selected publisher. The

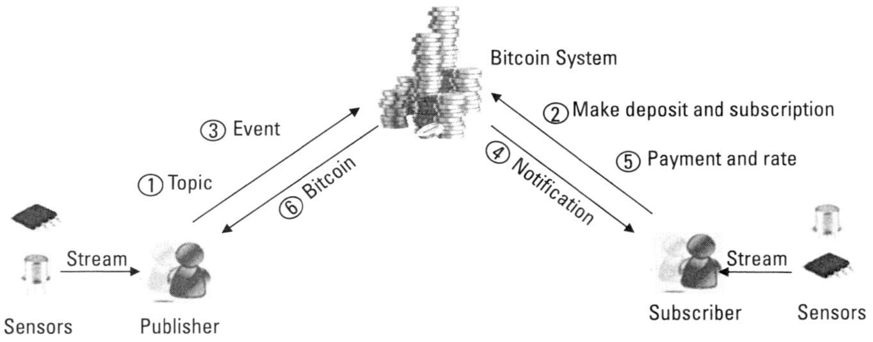

Figure 11.2 Trade management using the SPS model.

third step is that, on receiving the subscription request, the publisher matches the topic and broadcasts the matching event in the Bitcoin network. To ensure data confidentiality, the seller encrypts the data using symmetric encryption and transmits the ciphertext to the subscriber. The fourth step is that the subscriber is notified about the new event and verifies the publisher and/or the data based on the history of interactions in the blockchain. The fifth step is that, if the subscriber decides to proceed, the subscriber deposits a particular amount of coin in Bitcoin where the exact value is defined based on the reputation of the publisher. An untrusted subscriber is required to deposit additional coin compared to trusted subscribers. In case a malicious activity is detected by the subscriber, the deposit value will not be refunded. The sixth step is that the publisher finally obtains Bitcoins and reputation value.

Blockchain-based marketplaces facilitate the management process of the trade, which involves storing the transaction's history in a transparent and immutable manner. In a data marketplace, it is critical to maintain a catalog of the available data types and match them properly with the potential buyers, which is outlined in the next section.

11.2.2 Distributed Data Catalogs

In a blockchain-based marketplace, the participants in the trade are connected and share information through blockchain, which eliminates the need for central controllers. The sellers advertise features of their data including price, data types, and reviews in the blockchain. The buyers can explore the chain to find proper data type and connect to the seller to conduct the trade. Blockchain enhances the reliability of the listing process as the data is replicated in all the participants in an immutable manner. As the listings are managed by individual sellers, the likelihood of errors is low given that malicious entities in the marketplace cannot alter the price or the preference of any listing to benefit other sellers.

The authors in [3] introduced a method that enables the buyers to search available data types in blockchain and find a matching seller. Each seller initially creates a detailed product description that includes the data type, seller identification, price, and IP address. The seller then stores the file description in JSON format in a distributed file storage (DFS) framework such as IPFS or Storj. The DFS returns a storage identifier that points to the particular data stored in the storage and facilitates data retrieval. The seller then stores the storage identifier and product metadata in the blockchain through a product smart contract to manifest the product registration process. The product smart contract implements a data structure to store the protocol type and the hash index for each of the data products and functions to store product information on the blockchain. Blockchain functions as a product catalog that can be searched by buyers to find suitable data products.

The authors in [4] implemented an IoT data marketplace application as a smart contract targeting non-real-time and noncritical IoT applications. The IoT data marketplace provides various application binary interfaces (ABIs) to register users, add/query devices, and transfer data and defined data structures to record payload and user information on the blockchain. Only authorized nodes can push data to the system; thus, each vendor creates a new registry on the blockchain and declares the address of its devices. The end-to-end interaction of the data vendor, buyer, and IoT data marketplace smart contract can be explained as follows. IoT data generated from devices is uploaded from gateway to the Swarm file system in chunks in an encrypted form. The IoT devices use a hierarchically deterministic method to create new symmetric encryption keys for each upload. The vendor possesses the master keys of their gateway device. The data is recorded in a payload structure including the identifier of the device that uploads the data, the name of the encryption scheme (DES, AES), the key index, file handle, time stamp, and geolocation. Once the data is stored, the Swarm client returns a file handle, a unique identifier, and the address of the data to the gateway. The platform provides a flexible querying mechanism for data consumers who can query sensor data sets and receive a list of vendors who own relevant datasets. A consumer requesting more descriptive details such as time stamp, geolocation, or schema of the sensor data from a selected vendor shall pay the data owner for the requested data. On receiving the request, the selected vendor calculates the symmetric key that is used to encrypt that particular Swarm file by using that device's master key and key index. The vendor creates a transaction containing the symmetric key that is encrypted using consumer's PK^+. The consumer decrypts the symmetric key using the corresponding private key and then decrypts the Swarm file using the symmetric key.

The listing services allow the data producers to advertise their data and also allow the data consumers to search and retrieve particular data type. However, as outlined in Section 11.1, the data marketplace involves design decisions that are beyond data-listing services. The next section discusses marketplace architectures.

11.2.3 Hybrid Centralized-Decentralized Architectures

The existing data marketplace solutions can be categorized as centralized where a central controller manages all aspects of the trade and decentralized where the core functionalities of the marketplace are distributed between the participants. As outlined in Section 11.1, the central solutions introduce privacy, security, and single-point-of-failure challenges. A purely distributed method (i.e., blockchain-based method) also becomes challenging as storing all the indexing and management information in all the participants increases the storage and thus increases the cost of managing the chain (see Chapter 5). A blockchain-agnostic method using a hybrid approach (i.e., using both less critical centralized and new decentralized elements) creates an effective and realistic solution for data marketplaces. Hybrid approaches can either rely on brokers to manage trades, or simply rely on the central server to manage trading.

The authors in [5] introduced a broker-based architectural design for the monetization of IoT data using Ethereum smart contracts and the message queuing telemetry transport (MQTT) broker. The MQTT broker is a TTP that is hosted in the cloud to improve reliability, scalability, availability, and accessibility of the framework with low latency. The owner of the IoT device creates a smart contract for the device that facilitates money transfer, data rates and usage time auditing, issuance of access tokens, and log of the events. The blockchain stores the MQTT broker address and a list of topics provided by the device. Consumers who are interested in using and paying for specific IoT data make a deposit to the smart contract, which returns the broker address, a unique access token, and access duration. The access token is the hash of the concatenation of the owner and consumer's address, the topic ID, and the total number of accesses through the system. The data generated by the IoT device is collected and aggregated by the MQTT broker, which authenticates the consumer off-chain based on the token issued by the smart contract. Once authenticated by the broker, the customer will be given access to the IoT data through an MQTT connection for the specified period of time or until the customer disconnects.

Hermes was introduced as a brokerless data marketplace [7], an HTTP server-based platform for trading sensor data that employs distributed ledger technology as the underlying communication framework to enhance the security of the framework. A high-level view of the framework is shown in Figure 11.3. The sellers share metadata in the marketplace about the type of data that they are willing to stream and sell. The centralized marketplace server does not store or have any access to the actual data being streamed and only facilitates

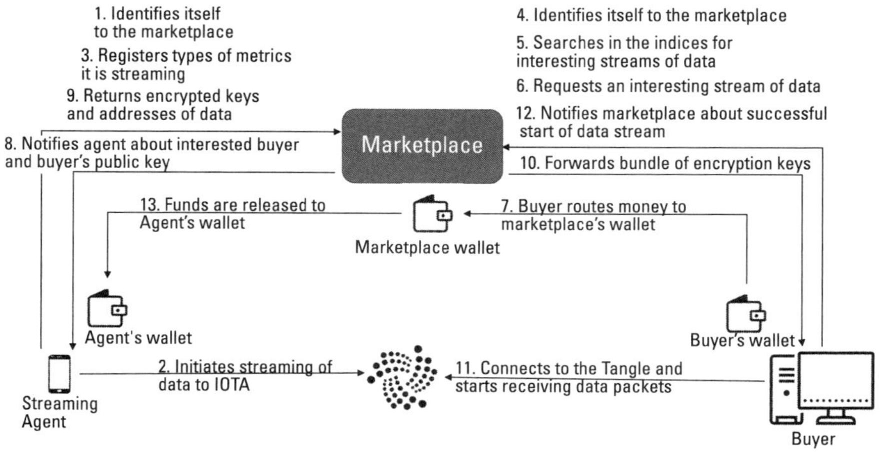

Figure 11.3 A high-level view of the framework introduced in [7].

the trading of streams of data by providing APIs to register or deregister users, post queries or offers, and raise disputes. Potential sellers can post advertisements in the central server. The IoTA is employed as the underlying blockchain framework. The buyer must first deposit the money in the marketplace wallet. Then the PK^+ of the buyer is forwarded to the seller along with the details of the purchase request. The seller packages the decryption keys of the stream and the transaction addresses, signs it with the corresponding private key, encrypts it with the buyer's PK^+, and sends it back to the marketplace to be forwarded to the buyer. The funds are released to the seller only when the bundle of data expires. This process allows dispute resolution in case the data is modified or the seller does not share the promised data.

The hybrid method is a combination of centralized and decentralized models that facilitates the functionality of data marketplace by reducing overheads and increasing trust and security. In the next section, we discuss decentralized data storage and managing access to the data, which is critical to a decentralized marketplace.

11.2.4 Decentralized Data Storage and Access Mechanisms

An access control mechanism enables the data owners to manage who can access their data and for how long. The access mechanism separates the data exchange into two parts: raw data exchange, where the participants share the raw data, and access right exchange, where the permission to access particular data is exchanged between participants.

The authors in [8] introduced Sash, which employs smart contracts to handle access control policies and evaluate access requests of buyers. Data

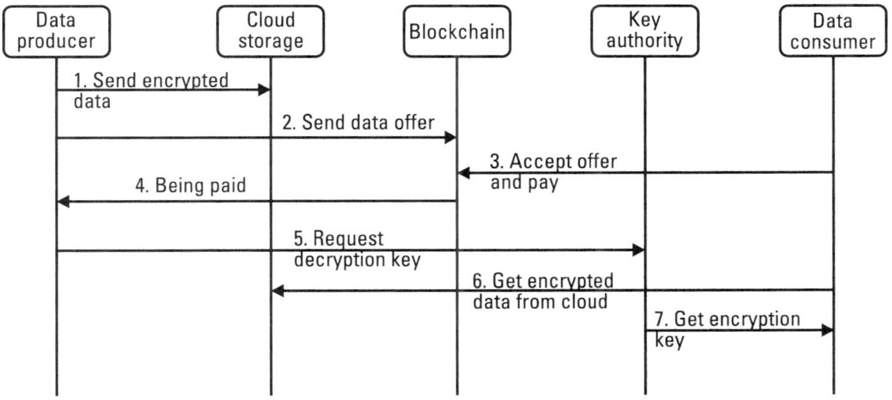

Figure 11.4 The process of prefix encryption.

owners store their data in an off-chain storage. Next, they advertise it to the blockchain participants by calling an "offer" function in the smart contract. The smart contract manages the access request to the data that is based on the predefined policies by the data owner. To address the malicious activity of untrusted storage providers, a data-sharing scheme is presented that employs prefix encryption as shown in Figure 11.5. The secret keys are distributed through a key distribution authority that holds the master secret key. When the key authority receives a request to access a given prefix, it uses the Access Control List (ACL) to decide whether to grant or deny the request. If the request is granted, the authority computes the secret decryption key and securely transfers that key to the requestor.

The authors in [9] advocated a data-centric design for IoT that abstracts away the location of data and separates access control from the data plane. The former is realized with a public blockchain and the latter is realized with a distributed data storage. IoT data streams are abstracted into data chunks and stored using a distributed hash table (DHT) (see Chapter 2), which acts as a private key-value data store interface. The DHT serves as a scalable, self-managing storage with high availability, while the blockchain is used to store the access permissions securely. A signed transaction contains the data owner, data readers, the corresponding data stream, and some additional metadata. Access rights are granted per data stream and are limited in time, as expressed in the number of chunks. To retrieve data, the responsible DHT node first checks the blockchain for access rights. A malicious DHT node may share users' data without permission. However, the impact of this action is limited since data is encrypted; each node holds a small random fraction of a data stream due to the nature of DHTs, and access right violation is detectable. The platform provides

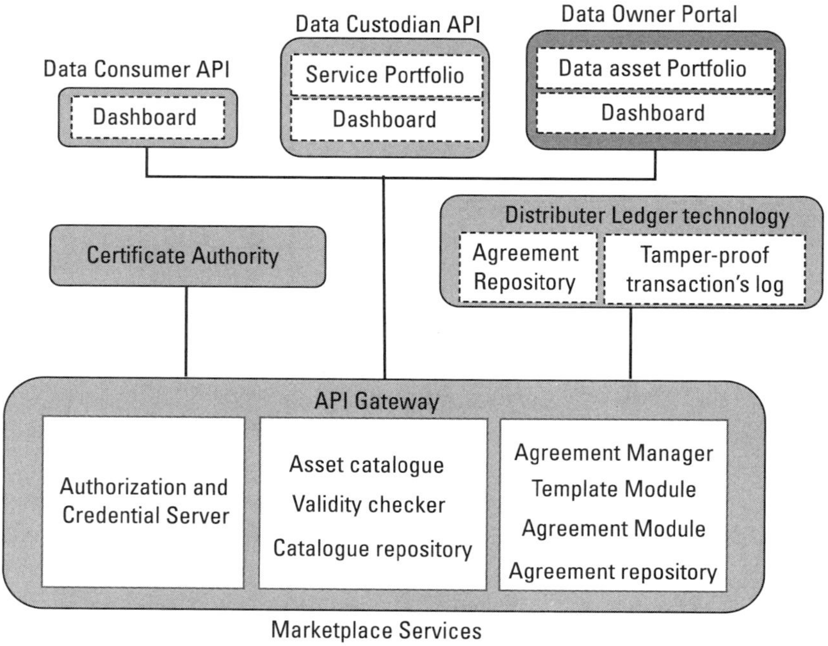

Figure 11.5 An overview of the framework in [10].

a reliable storage and distribution of data streams while empowering the data owner with fine-granular access control.

Access control management is critical to ensure that the user has control over the stored data and thus protect their privacy. To access data, the seller and buyer must reach an agreement, discussed in the next section.

11.2.5 Agreement Instantiation

An agreement instantiation approach manages the agreement life cycle and ensures the agreement's integrity, non-repudiation of the participants, and authentication.

The authors in [10] introduced an agreement framework that relies on a data marketplace used as a TTP handling all transactions among all stakeholders of the ecosystem, including data owners, data custodians, and data consumers. Blockchain is used to provide trust, transparency, and fairness in a multi-stakeholder ecosystem that enables storing the agreements established in the data marketplace as well as monitoring of the activities within the ecosystem. Data custodians are responsible for collecting data from data owner and transferring data assets to the data consumer. As shown in Figure 11.5, the framework consists of a set of data-related and asset discovery services, the agreement management, and the authorization server for agreement fulfillment. The data owners broadcast an agreement in the blockchain that describes acceptable terms and conditions for that particular owner. The data custodians broadcast a service-level agreement based on negotiated terms with data owner and data asset terms and conditions. The marketplace creates a transaction that contains the agreements. The involved parties validate the transactions and store the corresponding hash in the blockchain. The data and service assets are then saved in the service repository with the link to their respective agreement's hash.

This section has discussed existing IoT data marketplaces that leverage blockchain technology. However, most of the existing works have not considered the challenges in real-time data streaming from IoT devices. In particular, they have not considered the energy level of the sensor devices that participate in the marketplace while energy consumption is a critical factor in IoT. In [11], we introduced a new framework for trading real-time IoT data streaming from resource-constrained IoT devices known as the Autonomous IoT Data Marketplace (AIDM) Framework. The AIDM addresses the real-time data streaming requirement from the IoT devices by involving the battery status of the seller's device in the marketplace data-sharing process. The AIDM introduces several key features including an agreement instantiation, automatic execution of the agreement, and novel pricing and rating mechanisms. The next section discusses the AIDM in detail.

11.3 The AIDM Framework Using Smart Contracts

In this section, we outline the AIDM framework in detail. The AIDM is a hybrid data marketplace that uses facilitators to coordinate trading between data providers and consumers. The AIDM uses an agreement instantiation approach that provides an efficient data-sharing/trading platform and automates transaction workflow. The AIDM leverages smart contracts to provide fundamental marketplace functionalities including convenient estimation of data price based on the market dynamics and transparent assessment of the reputation of participating entities established through their trading history. The AIDM supports real-time data streaming from resource-constrained IoT devices using an optimization mechanism for the selection and allocation of buyers' demands on sellers' devices. We first provide a motivating use case for the AIDM in Section 11.3.1, followed by high-level overview of the AIDM in Section 11.3.2. We then discuss the major components of the framework in Section 11.3.3. The details of the optimization problem are discussed in Section 11.3.4. Section 11.3.5 covers the details of smart contracts to realize the different components of the marketplace.

11.3.1 Motivating Use Case

Consider an application developer providing location-based services by purchasing real-time location data from multiple sellers. The developer specifies his or her desired data quality in terms of accuracy and latency, sampling interval, and duration. These parameters directly impact the battery consumption of the seller's device. Existing techniques for improving power consumption of IoT devices usually compromise data quality for energy efficiency. High-quality data (high accuracy and frequency, low latency) usually consumes more energy. For instance, the positioning modality that provides the most accurate location information (GPS: 6m, Assisted GPS: 60m, Cell-ID/WiFi Positioning System: 1,600m) consumes the most power (GPS: high, AGPS: medium, Cell-ID/WPS: low) [9].

Similarly, when an IoT sensor works at a low sampling interval and for a longer duration, it imposes a heavy workload on different components (processor, network, storage, and sensors) of the device, draining the battery rapidly. As an example, if a smartphone owner sells his or her location data by sending frequent (low sampling interval) data, the phone battery depletes more quickly. This degrades the smartphone user experience and impacts the utility of other buyers with whom the seller has already made an agreement. However, if a low data rate setting is applied, then the data quality is reduced and the buyer's utility is degraded. Due to the costs associated with continuously sensing and transmitting data, a seller has to make an optimal decision to serve buyers' demands in real time based on the current residual energy and capabilities of the devices. Thus, it is vital to find a solution that maximizes the participants' utilities while taking into account the limited capabilities and constraints for creating a sustainable marketplace. With this use case in mind, we present the AIDM framework next.

11.3.2 Overview

The AIDM involves three actors: sellers, buyers, and facilitators. The network architecture is shown in Figure 11.6. A facilitator is a trusted entity and resource available entity whose motivation is to receive incentives in return for its service. The facilitators serve the following purpose and benefits in our system:

- *Scalability:* To enhance scalability, we leverage on the concept of geographically distributed facilitators that are interconnected in a P2P manner, thus forming an overlay network spanning the entire globe.
- *Catalog:* The facilitators maintain the list of the sellers' devices and the buyers' demands that belong to their service areas using distributed data catalog approach as explained in Section 11.2.2. The participants must

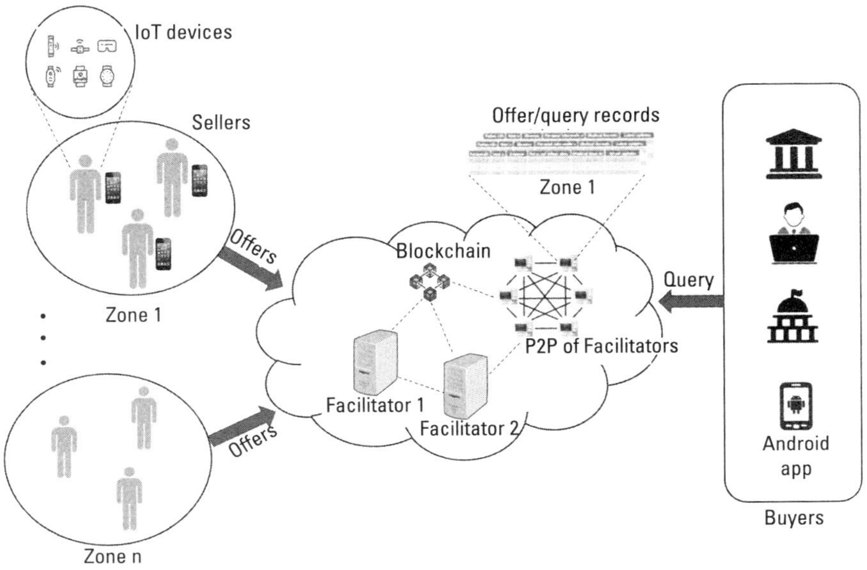

Figure 11.6 Network architecture of the AIDM.

register with a facilitator that is nearest to their geographical location and create a profile. A facilitator receives data queries from buyers and metadata for service offerings from sellers.

- *Search and match:* The facilitator uses a discovery and selection algorithm, an example is illustrated in [12] to match the service offerings and demands based on the criteria specified by both parties, such as location, data types, and budget pricing. Data streaming requires buyers to define quantitative parameters such as sampling interval, streaming duration, and data quality along with contextual parameters such as data type, location, and seller's rating. The facilitator matches selected datasets to the contextual parameters of the buyer's demand. The selection based on quantitative requirements needs to consider the battery availability on the seller's devices that is a dynamic parameter. To avoid the network overhead related to periodically notifying facilitators of the sellers' device battery status, the selection based on quantitative requirements is done locally at the seller's end using the optimization module. Its objective is to select and allocate the buyer's demands on the seller's devices under battery constraint such that the seller's revenue is maximized, which is explained in detail in Section 11.3.3.

Next, the facilitator creates the list of potentially matching buyers and sends it to all the identified sellers. This matching may involve participants in different zones associated with different facilitators, which highlights the demand for the facilitators to have a global view of the marketplace.

- *Dispute resolution:* At the end of data trading, both the buyer and seller submit the cumulative number of data samples transferred for the settlement purpose. The seller and buyer are untrusted entities which may behave maliciously. Such malicious behavior can be detected by the other party involved in the same trade and then is reported to the facilitator, who functions as a judge to resolve the dispute. The facilitator relies on the claims and evidence information provided by either side of the trade to solve the dispute. Based on the trading history of a user, a reputation score is recorded in the blockchain that signifies the probability that the user will behave well in the next transaction. In the case of detecting misbehavior using proofs, the facilitator penalizes the responsible party and reduces its reputation score and restrains the party from participating in the marketplace for a certain time period. If none of the parties is at fault and the dispute happens due to other reasons, such as channel congestion, the facilitator rewards both entities by increasing their reputation scores for being honest.

11.3.3 Main Components

Our marketplace framework supports the following main components:

- *Optimization-based selection and allocation:* The first step of discovering and matching of offers and queries is based on the contextual requirement of the demands (data type, location, and seller's rating) and is actioned by the facilitator as explained earlier in Section 11.3.1. The second step is actioned at the seller, which involves optimization-based selection using quantitative requirements (sampling interval, quality, duration) of buyers' demands.
- *Agreement framework:* The data streams of IoT devices require efficient, automated mechanisms to create legally binding trade agreements including payment arrangements and to enforce such agreements throughout data transmission. Using smart contracts as an agreement instantiation, sellers can specify their own terms and conditions that allow them to distribute their product details (e.g., price) and all other content, such as subscription detail, data quality, and data usage, in the way that best suits their needs. It also ensures that the participants' be-

havior automatically conforms to the terms of agreements. Moreover, smart contracts enable fine-grained per-user and per-data stream conditions to be formalized. The agreement framework consists of two smart contracts known as data subscription contract (DSC) to record subscription detail and register contract to record DSC detail explained in Sections 11.3.3.1 and 11.3.3.2, respectively.

- *Pricing model:* We consider dynamic pricing, which depends on market conditions. The AIDM thus supports a competition-based price model that enables a simple and dynamic evaluation of IoT data price using the pricing contract (discussed in Section 11.3.3.3) as compared with conventional high-resource-consuming methods.
- *Rating mechanism:* It is important for all the parties to rate each other based on their trading experience to establish a trustworthy and reliable framework. Our rating mechanism provides resilience to manipulation of reputation score from various attacks such as ballot stuffing (the user gives false-negative feedback to one entity to make him quickly lose his reputation), strike and rechange (the user builds up a good reputation by performing a number of low-value transactions well and then misbehaves in a very high-value one), and collusion (two users collude to influence a given reputation rating). The rating model utilizes the rating contract (Section 11.3.3.4) to record the trading history in blockchain and evaluates the reputation score of the actors to ensure reliability among trustless actors and add safeguards by tuning reputation scores based on the malicious behavior of dishonest actors.

The interactions of these components to achieve data trading are illustrated in Figure 11.7. The seller and buyer advertise offer and query, respectively, to the facilitator. The facilitator matches the offer and query and sends the potential list of demands to the seller. The seller then uses the optimization module to select demands based on the available battery of his IoT devices. Subsequently, the seller and buyer come to an agreement. A DSC is deployed in the blockchain and its address is registered using the register contract. The seller adds the subscription details in the blockchain using the DSC followed by an update of the price using price contract. The data is transferred off-chain from seller to buyer over the TCP connection. On completion of data transfer, settlement is done by submitting the feedback. At the same time, the payment is released to seller and invoice is generated for the buyer. This is followed by an update of the reputation score of both the actors via rating contract.

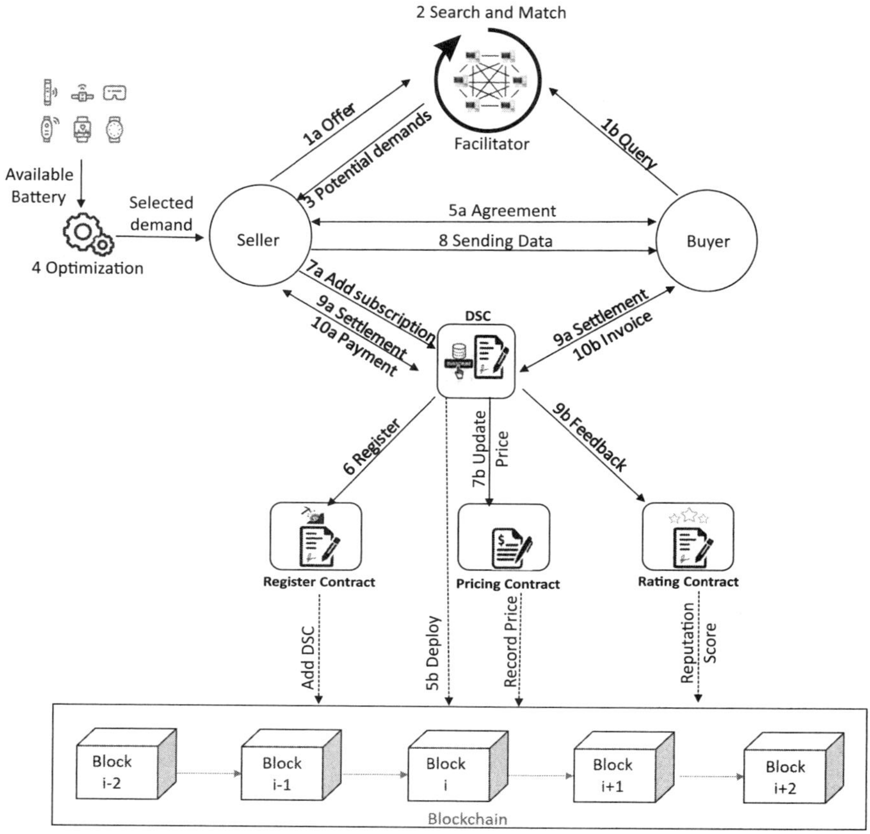

Figure 11.7 Interactions of components in the AIDM framework.

11.3.4 Optimization-Based Selection and Allocation

In order to address the need for streaming real-time data from resource-constrained IoT devices to the buyers, the AIDM proposes an optimization module to allocate devices of the seller to a selected subset of the demands in a way that the seller's revenue is maximized. The selection and allocation of demands should meet the quantitative requirements of the selected buyers and ensure that the battery capacity of the devices is sufficient to fulfill the assigned demands completely without any interruptions. This is related to the CPS persistence challenge. This optimization problem is defined as the Energy-aware Demand Selection and Allocation (EDSA).

The EDSA can be formulated as an Integer Linear Program (ILP) to find the optimal selection of demands of the buyers to maximize the revenue as given in (11.1) through (11.4). The formulation presented uses the following variables, categorized into two types:

1. *Seller's parameter:* A seller owns D resource-constrained devices with battery levels B_i, for $i = 1, 2, \ldots, D$. Each device provides a set of data types S. Without loss of generality, let Q_{ij} be the highest quality threshold that a *device* i can meet for a *data type* j. The unit data price $P_{ij}(q^k)$ is defined on the basis of the *device* i, *data type* j and the quality demand q^k of the *buyer* k. The energy consumption E_{ij}^k of *device* i is the energy required for sensing, processing, and transmitting *data type* j to the *buyer* k.
2. *Buyer's parameters:* There are a total of C potential buyers, where each buyer may have multiple demands with different requirement of data types. A *buyer k's* demands consists of *data type* j, and the duration of subscription is d_j^k with a sampling interval of s_j^k. Hence, $N_j^k = (d_j^k / s_j^k)$ is the total number of samples of *data type* j demanded by *buyer* k.

$$\underset{x_{ij}^k \epsilon \{0,1\}}{Maximize} \sum_i^D \sum_j^S \sum_k^C x_{ij}^k N_j^k P_{ij}\left(q^k\right) \tag{11.1}$$

subject to

$$\sum_k^C \sum_j^S x_{ij}^k E_{ij}^k \leq B_i, \ \forall i \tag{11.2}$$

$$\sum_i^D x_{ij}^k \leq 1, \ \forall k, \forall j \tag{11.3}$$

$$x_{ij}^k q^k \leq Q_{ij}, \ \forall i, \forall j, \forall k \tag{11.4}$$

The integer linear programming formulation outputs a selection of demands that optimally maximizes the total revenue while meeting the battery, quality, and allocation constraints. The objective function is the total revenue generated by the demands that are selected and allocated to an IoT device of the seller.

The above problem can be mapped to a Multiple Knapsack Problem with Assignment Restriction (MKP-AR). The MKP-AR is defined as follows. Its input is a bipartite graph $G = (X, Y, E)$ with a set of edges E between X and Y. The vertices of $X = \{x_1, x_2, \ldots, x_m\}$ correspond to knapsacks (sellers' devices), and the vertices of $Y = \{y_1, y_2, \ldots, y_n\}$ correspond to items (demands) where m

is the total number of sellers' devices and n is the total number of demands. E is defined between (x, y) if quality Q_i offered by *device* x_i is greater than the quality q_j demanded in *demand* y_j. Item $y \in Y$ is assignable to knapsack $x \in X$ only if $(x, y) \in E$ (assignment restriction). For each knapsack $x_i \in X$, its capacity c_i (battery capacity) is associated. For each item $y_j \in Y$, the profit (price) and the weight (power consumption) of y_j, denoted by p_j and w_j, respectively, are associated. A feasible solution of MKP-AR is an assignment of items to knapsacks such that, for each x_i, the total size of assigned items to knapsack x_i is at most c_i. The goal of the MKP-AR is to maximize the total profit of assigned items.

11.3.5 Marketplace Components Using Smart Contracts

The AIDM realizes key components of the marketplace by leveraging smart contracts. There are two types of accounts in the blockchain: external account and smart contract account. Once the actors register with the marketplace, they receive a dynamic key pair PK^+/PK^- and become the external account holder. The smart contract is compiled into bytecode and deployed in the blockchain with unique addresses that can be called by any external account holder using a transaction or by another contract using messaging. To call the function within smart contract, the ABI and smart contract address must be specified. Agreement details, participants' ratings, and data price are recorded in a transparent, secure, efficient, and automated way in the blockchain for future use. The methods provided by each contract and the transaction detail to execute them are given in detail below.

11.3.5.1 DSC

All the agreed and negotiated terms are encoded, compiled, and deployed in the blockchain as a smart contract known as the DSC. There is a DSC for each seller-buyer pair and its state corresponds to the subscription details like subscription ID, seller ID, buyer ID, device ID, data type, start time, periodicity, duration, quality score (*QS*), risk score (*RS*), total cost, payment granularity, transaction status, and negotiation information. Device ID is the hash of the device H/W serial number to ensure that the seller's claim of owning the device is correct and the information is stored in a hash form for privacy purposes. The DSC provides various ABIs to manage the subscription list and implement the data trading such as:

- *subscriptionAdd* to add a new subscription request to the existing subscription list;
- *SubscriptionInfo* to receive the information required for subscription and return the subscription result;

- *SubscriptionSettlement* to perform settlement by releasing payment and generating invoice;
- *SubscriptionDelete* to delete the subscription entry from the subscription table.

11.3.5.2 Register Contract

Register contract is a multisig contract that requires both involved parties to sign the contract and is used to deploy and register the DSC. It maintains a contract look-up table that records the details of all the DSCs consisting of the following fields: contract ID, creator ID, seller public key, buyer public key, DSC address, and ABI's name. The register contract uses multiple functions to manage the look-up table including *contractCreate* to add a newly deployed DSC in the contract look-up table and *contractRemove* to delete the contract entry from the look-up table using the contract ID, which will subsequently perform the *SelfDestruct* operation of DSC contract to remove the code and storage of the DSC from the blockchain, such that the DSC can no longer be available. The interactions of DSC and register contract are depicted in Figure 11.8.

11.3.5.3 Pricing Contract

A pricing contract as shown in Figure 11.9 is deployed for each data type available in the marketplace. The pricing contract maintains a ledger with the following fields: time stamp, data type identifier, price, quality score, and risk score. This contract is invoked by the message sent from the DSC whenever a new subscription is added in the agreement to record the price of the data type. The values stored in the ledger are used to calculate the base price offered by the other sellers in the market.

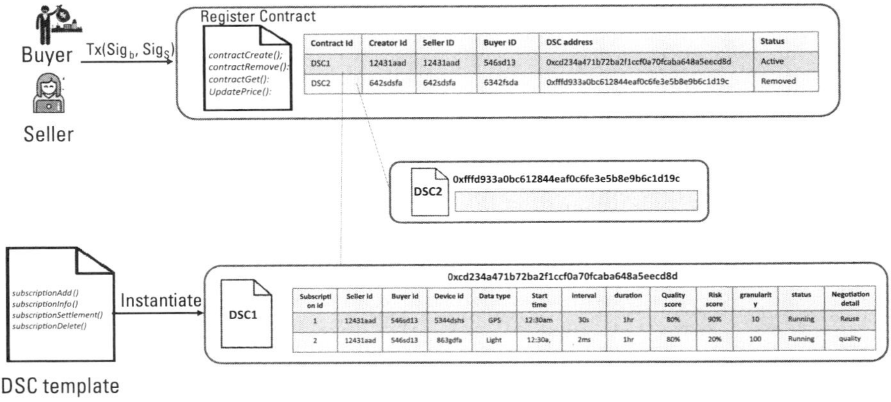

Figure 11.8 DSC and Register contract.

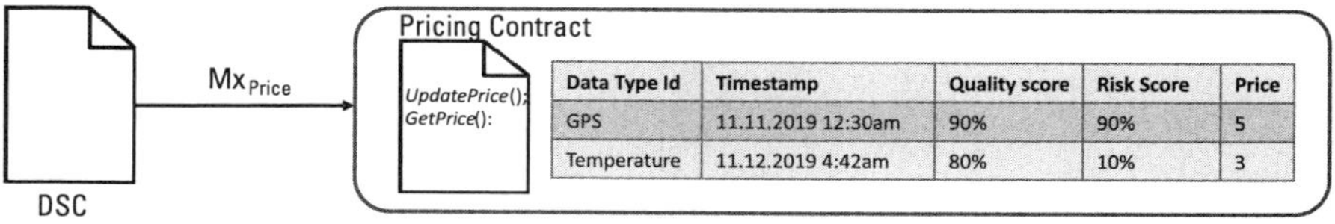

Figure 11.9 Pricing contract.

We have adopted a competition-based pricing in our framework because a data marketplace is a highly competitive market with a large number of sellers interested in offering their IoT device data in return for incentives. In competition-based pricing, the seller makes use of competitors' prices for the same data type as basis in setting a price. This strategy does not require any complex computations and varies depending on the market. The seller can follow the market going rate of a certain data type as charged by other sellers selling the same data type. However, price is not a major differentiating factor to draw the buyer's attention, and, thus, the seller must provide value-added features that depend on the quality of the data and the magnitude of privacy risk associated with a particular data type. In addition to this, smart contract-related execution fees can also be divided between the seller and the buyer as per the negotiated terms. The price model uses these parameters in evaluating the price of the data as given in (11.5).

$$Price = (1 + QS + RS)P_{base} + \beta P_{exe} \tag{11.5}$$

where *QS* is quality score, *RS* is privacy risk score, *NR* is negotiation rate, P_{base} is the base price, P_{exe} is the execution price, and β is the agreed-upon share of P_{exe}. These parameters are explained in detail as given below.

- *Base price* (P_{base}): The competitor's price is used as a benchmark rather than pricing based on buyer demands or other mechanisms. It corresponds to the base price of the data type, which is a fixed component and non-negotiable. Using the price detail recorded in the ledger, the price index $P^i_{d,t}$ is calculated and then it is averaged for the number of successful trades for a given interval. For the total value of the *N* transactions for *d* data during a given interval of time *t*, the base price is given as in (11.6) and (11.7):

$$P^i_{d,t} = \frac{Price^i_{d,t}}{QS^i_{d,t} + RS^i_{d,t} + 1} \tag{11.6}$$

$$P_{base}^{d} = \frac{\sum_{i=1}^{N} P_{d,t}^{i}}{N} \quad t\epsilon T \tag{11.7}$$

- *Quality score* (*QS*)*:* Quality score is calculated based on the buyer's quality demand and preferences and is the weighted average of the quality level desired by the buyer with the weights dependent on his or her preferences. The contract calculates the quality price as the quality score percentage of base price.
- *Privacy risk score* (*RS*) *[13]:* The IoT data holds intrinsic value and sharing it exposes the data owner to potential damage represented as risk, which can lead to a number of privacy harms. Therefore, a seller can categorize each data type into different harm types (*T*) and the impact of risk represented as consequences (*C*) as shown in Figure 11.10(a). *T* depends on whether the privacy harm of the data type is distorting the data owner's identity or reputation, revealing information about the data owner that was never meant to be exposed or intruding upon the data owner's life. *C* depends on the impact of the risk, which ranges from insignificant ($C = 1$) to critical ($C = 5$). Based upon harm type and risk consequences, the risk level of a data type is estimated that is finally used to calculate the risk score as shown in Figure 11.10(b).

11.3.5.4 Rating Contract

In a marketplace, the actors should be trustworthy as economic interests are relevant. Therefore, reputation plays a key role in consumer-to-consumer (C2C) trading. Maintaining reputation scores is important in building trust, facilitating smooth transactions, and reducing risk. The reputation model considers the actor's trading history. The rating contract, depicted in Figure 11.11, maintains the reputation score of each actor in the marketplace, which signifies the probability that the actor will behave normally in the next transaction. On the contract completion, both the parties are required to submit their feedback. Post-settlement, the DSC sends a message to the rating contract containing feedback of both actors involved in the trade, which is used to calculate the overall reputation score RS_j given by (11.8).

$$\begin{aligned} RS_j &= \left((1-\alpha)RS'_j + \alpha F_j\right)e^{-\frac{c_{fail}}{c_{total}}} \\ \alpha &= (FC + TV)CA \end{aligned} \tag{11.8}$$

RS_j is the current reputation score of the *actor j* at time *t*. α is used to tune the latest feedback F_j and the last reputation score RS'_j of the *actor j* at time $t-1$.

Risk Consequence (C)

	Risk Level (C x T)	Insignificant (C=1)	Minor (C=2)	Moderate (C=3)	Major (C=4)	Critical (C=5)
Harm Type (T)	Distorting (T=1)	LOW (1)	LOW (2)	LOW (3)	MEDIUM (4)	MEDIUM (5)
	Revealing (T=2)	LOW (2)	MEDIUM (4)	MEDIUM (6)	HIGH (8)	EXTREME (10)
	Intruding (T=3)	LOW (3)	MEDIUM (6)	HIGH (9)	EXTREME (12)	ABSOLUTE (15)

(a)

Example of Data Set A

Type (T)	Consequence (C)	Risk Level (C x T)
Distorting (T=1)	Major (C=4)	4
Revealing (T=2)	Critical (C=5)	10
Intruding (T=3)	Minor (C=2)	6
Risk Score		**20/30 (67%)**

(b)

Figure 11.10 (a) Risk matrix, and (b) risk score example.

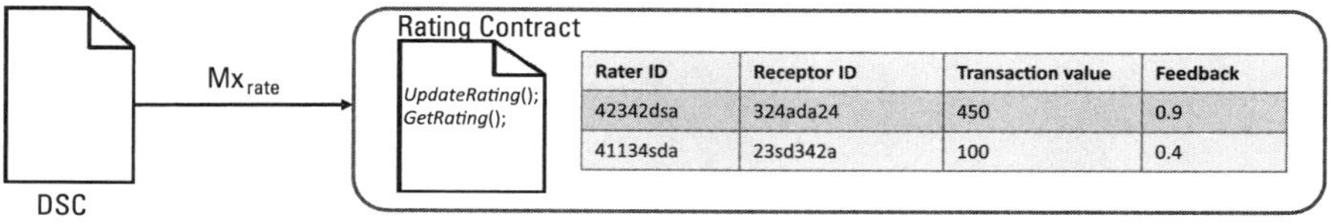

Figure 11.11 Rating contract.

α considers the feedback credibility (*FC*), transaction value (*TV*), and the collusive activity (*CA*). These parameters are suitable for successful contracts only. We also consider the case when the agreement is violated and is not fulfilled by seller. In the case of a failed contract, the violation factor (*VF*) is used to decrease the reputation score. The exponential factor used in (11.8) reduces the effect of *VF* when the number of failed contracts is very low compared to the total number of contracts. The various parameters used to calculate the reputation score are explained below.

- *Feedback credibility* (*FC*): This is a way to consider the actor's reputation score when the actor is providing feedback to misuse his or her high reputation score. A feedback is said to be positive (F^+) if F_j is greater than 0.5 and negative (F^-) if it is less than 0.5. The first part in (11.9) is for differentiating dishonest feedback from the honest feedback by considering the reputation score (RS_i) of the *actor i*. The second part is to prevent a ballot-stuffing attack by preventing a dishonest user from giving false-negative feedback to maliciously modify another actor's reputation score.

$$FC = \frac{RS_i}{RS_i + RS_j} \cdot \left(\frac{F^+}{F^- + F^+} \right) \qquad (11.9)$$

- *Transaction value* (*TV*): This parameter is the ratio of the current transaction value (V_c) to the maximum transaction value (V_{max}) of all the trades happening between two users as given in (11.10). Tuning the feedback proportional to *TV* slightly increases the reputation score for low-valued transactions and heavily increases it for high-valued transactions. Hence, preventing a dishonest user from performing a strike and recharge attack.

$$TV = \frac{V_C}{V_{max}} \qquad (11.10)$$

- *Collusive activity* (*CA*): A collusion factor is used to avoid the possibility of actors colluding with one other to increase each other's reputation scores. It is inversely proportional to the number of transactions $T_{i,j}$ made between *actor i* and *actor j*. *CA* is calculated using (11.11) where *a* is a non-negative factor and depends on the relevance of the feedback.

As the transaction numbers ($T_{i,j}$) between two actors increases, which implies collusion, *CA* tends to 0.

$$CA = \left(\frac{1}{T_{i,j}}\right)^{a} \tag{11.11}$$

- *Violation factor* (*VF*): The violation factor (*VF*) is a straightforward mechanism that is used to define the reliability of the seller to fulfill the contract without any interruption in the data trading due to resource-constrained IoT devices. The reputation score of the sellers drastically decreases if they fail to fulfill the agreement, which will ensure that the sellers actually deliver the contract to which they agreed. Therefore, *VF* holds the maximum weightage in calculating the reputation score for sellers and *VF* is 1 for buyers. During the agreement violation, α and F_j are 0. *VF* is calculated using failed contracts (C_{fail}) and the total number of contracts (C_{total}) as given in (11.12), and the reputation score in the case of contract failure is calculated using (11.13).

$$VF = 2 - 2^{\frac{c_{fail}}{c_{total}}} \qquad (11.12\_$$

$$RS_j = RS'_j.VF \tag{11.13}$$

11.4 Critical Analysis

In this section, we provide a high-level analysis of the marketplace frameworks studied in this chapter. The results of the evaluation are summarized in Table 11.1.

11.5 Conclusions

In this chapter, we studied the existing blockchain-based solutions for realizing the different functionalities of a data marketplace. We categorized the existing methods in five main categories: trade transaction management, distributed data offering listings, hybrid architectures, decentralized data storage and access mechanisms, and agreement instantiations. We studied a novel solution to address the outlined challenges as the AIDM framework. In the AIDM, func-

Table 11.1
An Analysis of the Existing Blockchain-Based Solutions in the Marketplace

Blockchain Method	Usability	Advantage	Disadvantage
Trade transaction management	Stores trade logs	Auditing and transparency	Does not scale
Distributed data catalog	Record data offerings	Reliability in listing seller's offering	Limited offered services
Hybrid centralized-decentralized approach	Pays the price involved in the trade	High feasibility	Centralization issues
Decentralized data storage	Maintains access policy	Owner controlled data	High redundancy
Agreement instantiation	Manages agreement details	Non-repudiation	Specific functionality of marketplace
AIDM	Records subscription details, data pricing trend, participant's behaviour	Autonomous, automatic decision, full-fledged with key functionalities	Requires two separate blockchains

tionalities of four smart contracts (data subscription contract, register contract, pricing contract, and rating contract) are discussed and their associated methods to achieve the trustful automated data trading by eliminating the third-party risk. The framework also contributes in providing a simple and dynamic way of pricing the data based on the competitors' prices in the market and encourages sellers to sell higher-quality data and value-added features in return of greater incentives. In order to provide resilience to various security attacks, a reputation mechanism based on the C2C rating model is presented that uses the trading history of entities in tuning the reputation score and penalizes them for being dishonest. The AIDM is underpinned by an EDSA mechanism that optimizes the revenue for the seller and fulfills the buyer's demands to facilitate real-time data streaming from resource constrained IoT devices.

The trading of data with IoT data marketplaces restricts transactions within the digital domain, which simplifies the trust management issues that we have seen in Chapters 8 to 10. However, the cyberphysical coupling in the data marketplace context involves the operational constraints of the IoT sensors, such as battery life (persistence), computational resources, and bandwidth. This dependency necessitates distributed algorithms for managing the multi-objective optimizations involving the IoT node operational constraints, as well as the profit of the sellers and the utility of the buyers. We have seen that blockchain can provide the information infrastructure for managing the involved trade-offs in a distributed manner.

This chapter concludes Part III of the book where we have explored how blockchain can support CPS applications that involve traceability and monitoring of physical and digital assets. In the next chapter, we revisit the key findings in the book and chart out the open research questions and future directions.

References

[1] Krishnamachari, B., J. Power, S. H. Kim, and C. Shahabi, IoT Marketplace: A Data and API Market for IoT Devices, University of Southern California, Available at: https://ms-bfile03.usc.edu/digitalmeasures/gerardpo/intellcont/USCIoTMarketplace_Jan152017-1.pdf.

[2] Zhao, Y., et al., "Secure Pub-Sub: Blockchain-Based Fair Payment with Reputation for Reliable Cyber Physical Systems," *IEEE Access*, Vol. 6, 2018, pp. 12295–12303.

[3] Ramachandran, G. S., R. Radhakrishnan, and B. Krishnamachari, "Towards a Decentralized Data Marketplace for Smart Cities," *2018 IEEE International Smart Cities Conference (ISC2)*, September 2018, pp. 1–8.

[4] Özyilmaz, K. R., M. Doan, and A. Yurdakul, "IDMoB: IoT Data Marketplace on Blockchain," *2018 Crypto Valley Conference on Blockchain Technology (CVCBT)*, June 2018, pp. 11–19.

[5] Suliman, A., et al., "Monetization of IoT Data Using Smart Contracts," *IET Networks*, Vol. 8, No. 1, 2018, pp. 32–37.

[7] Tzianos, P., G. Pipelidis, and N. Tsiamitros, "Hermes: An Open and Transparent Marketplace for IoT Sensor Data over Distributed Ledgers," *2019 IEEE International Conference on Blockchain and Cryptocurrency (ICBC)*, May 2019, pp. 167–170.

[8] Truong, H. T. T., et al., "Towards Secure and Decentralized Sharing of IoT Data," *2019 IEEE International Conference on Blockchain (Blockchain)*, July 2019, pp. 176–183.

[9] Shafagh, H., et al., "Towards Blockchain-Based Auditable Storage and Sharing of IoT Data," *Proceedings of the 2017 on Cloud Computing Security Workshop*, November 2017, pp. 45–50.

[10] Koul, A. S. N. A., and J. H. Morin, "Agreements Framework for Data Market Ecosystem," *Proceedings of the 13th Pre-ICIS Workshop on Information Security and Privacy*, Vol. 1, December 2018.

[11] Gupta, P., S. S. Kanhere, and R. Jurdak, "A Decentralized IoT Data Marketplace," *Proceedings of the 3rd Symposium on Distributed Ledger Technology*, Gold Coast, Australia, November 2018.

[12] Charpenay, V., et al., "Matching Offerings and Queries on an Internet of Things Marketplace," *European Semantic Web Conference*, June 2018, pp. 66–71.

[13] Molina, V., M. Kersten-Oertel, and T. Glatard, "A Conceptual Marketplace Model for IoT Generated Personal Data," *arXiv preprint arXiv:1907.03047*, 2019.

12

Open Research Questions and Future Directions

12.1 Concluding Remarks

This book has taken us on a journey in exploring blockchain as a solution to many of the challenges faced in CPS. We also explored how blockchain can support specific CPS-supported application areas. Figure 12.1 summarizes the topics we covered in the book. Here, we share our reflections on the prevalent trends we have observed.

CPS development and adoption have been driven by an increasing need for automation in modern economies. This need has fueled the exponential growth in IoT sensors to tens of billions, underpinning a broad range of applications from smart cities to smart grids. With this surge of CPS adoption, significant gaps relating to security, privacy, and trust were not given sufficient attention, which limits the confidence in and utility of these systems. Centralization of communication and trust through the conventional stovepipe model is a scaling limitation in these systems as well.

Many salient features offered by blockchain technology have motivated its use as a panacea to some of these challenges faced by CPS. The intuition for applying blockchain for these systems is its intrinsic decentralization, which mitigates the challenges around centralization and scale for CPS, and its security and privacy, which address a significant CPS gap. Throughout this book, we have focused on how blockchain can address specific CPS technical gaps and have

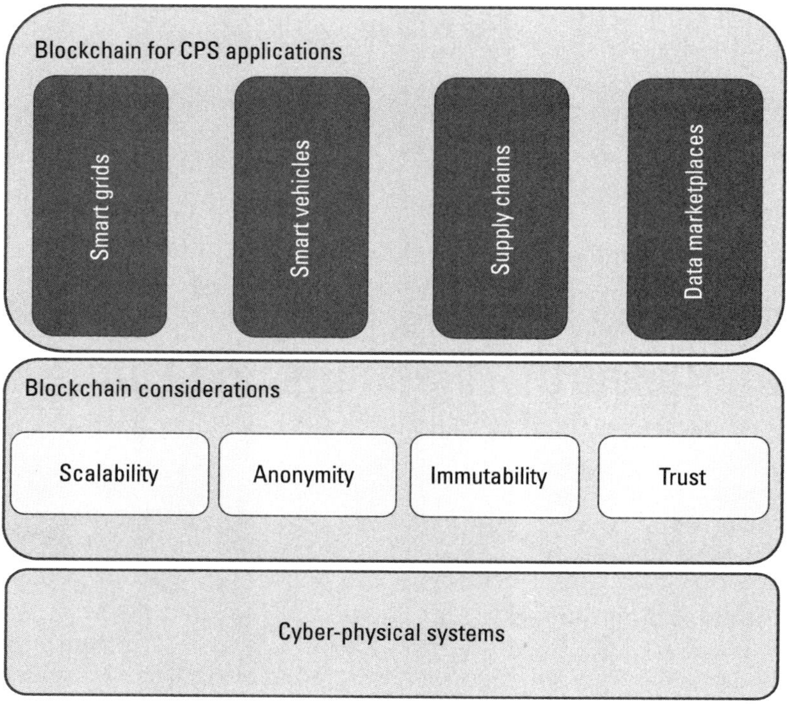

Figure 12.1 Overview of the topics that we covered throughout this book.

shown that blockchain can provide a trustworthy information infrastructure across several CPS applications.

We initially examined the suitability of core blockchain functions, such as consensus to CPS contexts. We found that most consensus mechanisms for cryptocurrency or financial applications are generally not well suited for CPS scenarios. In fact, the most promising lightweight blockchain approaches for CPS have adopted a design philosophy that directly targets CPS-specific design considerations, including limited computational resources, scalability, and interfacing of the digital and physical realms.

We next turn our attention to blockchain immutability, which is one of the technology's most attractive features in establishing data storage integrity. Recent data privacy regulations that are emerging across various jurisdictions, such as the EU GDPR, present a hurdle to the adoption of blockchain for storing privacy-sensitive data due to the persistence of this data on the ledger. We observed a growing trend of proposals to reconcile blockchain's immutability with these regulations that enforce the right to be forgotten. The most prominent of these proposals achieve a high degree of flexibility in managing what is stored on the blockchain and for how long, while maintaining blockchain consistency throughout any alterations to the stored content.

Our journey into trust and blockchain appeared paradoxical at first, as blockchain is aimed at providing consensus in the absence of trust among participants. In CPS, however, the externality of trust-related risks to the blockchain system itself has motivated exploration of this interplay between trust and blockchain. More specifically, blockchain cryptographically seals the content stored on the ledger, but it does not provide any guarantees on the identity of the participants that generated the content or the integrity of the content as it was observed from the physical domain. As a departure from the conventional self-contained blockchain systems, the solution to the trust problem is clearly in augmenting existing blockchain systems with trust and reputation constructs. Importantly, these constructs need to provide an integrated trust score that cuts across the different dimension relating to the entity that generates the data, the data acquisition integrity, in addition to the data storage integrity provided by blockchain.

The next step in our journey in this book was anonymity. With the significant increase in the number of CPS and IoT devices, which can create digital stories of physical events that often reveal private information, there is an increasing need for providing anonymity around these systems. At a high level, these risks are common with any digital systems, including the internet.

The challenge with blockchain is that the data persists indefinitely, as the ledger is by definition, immutable. The data persistence therefore creates long-term anonymity risks for the data generators. Adding to that challenge is that the anonymity vulnerabilities themselves are not known when the data is generated, but are discovered later and are able to exploit the data persistence. We saw an example of this scenario in using intertransaction times on blockchain-supported CPS to identify devices in a smart home that then allow an attacker to infer the occupant's activity. We expect there will be analogous vulnerabilities in other blockchain-supported CPS scenarios that will need to be identified and addressed.

It is important to note that there is an intrinsic trade-off between trust and anonymity. As in social relations, CPS participants can typically trust the identity, data, and behavior of other participants whose identity and past data and behavior have proven to be genuine and reliable. In other words, you trust who you know. This is in tension with the participant's need to maintain anonymity in many scenarios, including, for instance, in supply chains or energy trading networks. Providing anonymous verification of certain attributes of participants, for instance, through zero-knowledge proofs, is likely to be increasingly investigated as a potential solution for this challenge.

Our exploration of the four blockchain challenges for CPS, namely scalability, immutability, trust, and anonymity yielded guidelines, best practices, and open research areas that are application-agnostic. CPS applications that can benefit from blockchain also have their own specific context, challenges, and opportunities that we explored next. The first application area focused on energy trading. The penetration of two-way communication model and distributed energy sources (DES) have highlighted the need for a distributed information infrastructure to underpin the smart grid. Unique to smart grid applications is that they involve the management trading of pooled resources, fundamentally electrons, across a large set of distributed participants, which make traceability and auditability of transactions in the grid challenging, and provide new opportunities for blockchain solutions. We studied four applications of blockchain in smart grid. The first was energy trading. By penetration of DES, the end users can generate and inject the surplus energy to the smart grid. Blockchain enables the energy producers to trade the energy directly with end users without relying on TTP such as the grid manager. The second was data management. Smart grid is equipped with a large number of sensors and devices that produce a large volume of data. Blockchain is explored as a trusted database where users can store their data or/and authorize other parties to access their data. The third was demand side management. The distribution companies may sign a contract with customers that allows them to control a number of devices in the user site and control the load of such devices to balance the network load. Blockchain introduces auditability, security, and anonymity, which makes it an attractive solution for demand side management. The fourth was emission credit trading. Blockchain is applied for distributed carbon emission credits trading. Blockchain ensures the security and establishes trust among the participants.

The smart grid is evolving quickly, and emerging technologies such as sensing and blockchain can have major positive impact on its evolution. A nontechnical challenge that needs to be overcome is for governments and electricity markets to adapt their regulatory frameworks and operation and business models to these emerging technologies. For instance, grid operators have significant investments in the cables that provide connections on grids. As DES become more prevalent, customer reliance on this infrastructure will progressively diminish. However, operators may continue to benefit from their investments by adopting new business models that facilitate energy trading or distributed energy management through their infrastructure and by shifting to a more open and service-based model. Similarly, governments would need to embrace the new network architectures and their associated opportunities by adopting suitable regulations that support more participatory energy and carbon trading, which is expected to yield economic and environmental benefits.

Another interesting application of blockchain is smart vehicles, which involve high mobility, increasing wireless connectivity, and autonomy. The mobility of smart vehicles is particularly challenging for keeping track of relevant events. Smart vehicles are equipped with a broad range of sensors that introduce advanced vehicle functionalities and facilitate driving decisions by providing a better perception of the road. The highly connected nature of smart vehicles that facilitates these applications also introduces new challenges related to security, privacy, and liability. These challenges require targeted blockchain solutions for maximizing the technology's benefits in the smart vehicle space. Blockchain is explored as a framework to address the outlined challenges due to its salient features including decentralization, security, and transparency. We provided a comprehensive review of the existing blockchain-based frameworks in automotive industry and categorized those in four groups. The first was automotive network security, where securing communications and data exchanges between smart vehicles is critical. Blockchain transactions contain a hash of the exchanged data and are cryptographically signed, which makes it an attractive solution to address this challenge. Mobility can be managed through handovers among the blockchain miners. The second was trust and reputation management. Smart vehicles communicate and exchange data that may impact the driving decision of other vehicles; thus, it is critical to establish trust to the data itself and the data generator. Blockchain immutability makes it an attractive solution to establish trust and reputation, which increases the security of the framework. The third was privacy. The data of the smart vehicles contains privacy-sensitive information (e.g., the location of the vehicle), which makes privacy highly challenging. The anonymity offered by the blockchain can enhance the user privacy in smart vehicles. The fourth was vehicle forensics. The decision made by a vehicle is impacted by a number of sensors coming from different manufacturers, which complicates liability attribution when an accident occurs. Blockchain provides an auditable transparent framework and thus can serve as a common source of truth among the vehicular network participants to determine the liable party for an accident. We provided a risk analysis for the smart vehicles where we discovered the in-vehicle communications as the main source of threats. We introduced a blockchain-based framework to secure smart vehicles and address the liability challenge. The issue of liability attribution for vehicles is likely to grow in importance as vehicles become even more autonomous and connected. Recent fatal accidents involving autonomous vehicles highlighted the need for rapid and reliable methods to pinpoint what has gone wrong and who is to blame. There are obvious links here to the area of robot ethics that explores responsibilities in relation to the design and operation of autonomous systems. Keeping the relevant sensor data in silos with the major players in this space is detrimental to transparent liability attribution. Blockchain, along with relevant regulation changes from governments, is likely to play a key role in addressing this issue.

We next studied supply chain as an application of blockchain-based CPS. Supply chains share some features with energy networks in terms of asset trading among untrusted participants. Unlike energy markets, supply chains involve the trade of discrete physical goods. Supply chains have increasingly incorporated a wide range of sensors that capture environmental parameters to trace the production history of a commodity that includes origin, owners, manufacturing, and logistics. Supply chain involves parties that may potentially be located in different geographical locations, and, thus, various legislations may apply that further complicate supply chain process. The consumers demand traceability of the product origin to ensure the quality of a product. The physical nature of trades in supply chains raises challenges on cyberphysical data trust, which we discussed in Chapter 5. Blockchain is explored as an auditable and transparent solution to address the outlined challenges. Designing a blockchain-based supply chain involves fundamental requirements which include deciding on whether to use a public or private blockchain, identifying the responsibility of each party in the blockchain, the stage of the supply chain where blockchain can be incorporated, and identifying the type of data to be stored in the blockchain. We provided a comprehensive study on the state-of-the-art solutions that employ blockchain as a platform for supply chain management. Traceability is the fundamental feature of blockchain that makes it attractive for supply chain frameworks. In a blockchain-based system it is critical to ensure the authenticity of the data produced by the devices and stored in the blockchain. Supply chains are highly amenable to the benefits of blockchain coupled with CPS sensors, which can provide clear benefits to consumers and to food safety. Some of the barriers to wider adoption of these technologies in supply chains relate to market competition, where supply chain participants may be reluctant to reveal commercially sensitive information that may benefit their competitors. Protection of this information while delivering the traceability and auditability benefits is therefore key to the adoption of blockchain for CPS in supply chains. Another modulating factor will be the extent of consumer pressure for traceability of products, and government regulation in support of transparency, safety, and fair competition.

The final application of blockchain for CPS that we covered is IoT data marketplace. IoT devices produce a huge volume of data that is mainly stored in silos by the SPs and the data access is limited to the SP and the user. Given the significant value of the data of IoT devices, there is an increasing demand for marketplaces where the users can monetarize and trade their data. Unlike energy trading, where pooled electrons are traded, or supply chains, where discrete physical goods are traded, IoT data marketplaces transfer discrete digital assets (i.e., data) among participants. The restriction of traded assets to the digital realm avoids the challenges around entity or data trust, as trades can be verified digitally. However, IoT data marketplaces involve their own unique challenges, in terms of balancing the computational resources at IoT devices that are supplying traded data with the benefits for both data suppliers and consumers. Blockchain can underpin these trades and enables users to trade their data without relying on TTP and authorize nodes to access their data. User anonymity, traceability, persistence, user control, and distributed data catalogs are the key benefits of blockchain in the IoT data marketplace. We categorize the existing data marketplace solutions in five groups: (1) the trade transaction management, involving multiple steps in trading data including communication, negotiation, execution, settlement, and payment; (2) the distributed data catalog, listing data in blockchain so that the potential sellers can locate potential data and the corresponding data seller; (3) hybrid centralized-decentralized architectures, data marketplace frameworks where a combination of centralized and decentralized management is employed; (4) decentralized data storage and access mechanisms, storing data in a distributing way and authorizing users to access the data; and (5) agreement instantiations, managing the life cycle of an agreement between data seller and buyer that includes integrity, non-repudiation, and authentication. We introduced a comprehensive IoT data marketplace framework that eliminates reliance on TTP by introducing four smart contracts: subscription, registration, pricing, and rating.

A key feature of IoT data marketplaces is ensuring that the computational constraints of data suppliers, such as battery energy, processing capacity, and communication bandwidth, are considered in determining which data requests a supplier can service. This represents a multi-objective optimization problem that incorporates the computational constraints of data suppliers, requests from data consumers, and the need to maximize profits and value for the supplier and consumer, respectively. This can lead to real-time and dynamic IoT data marketplaces that are highly versatile. Blockchain's role here is to provide a decentralized information infrastructure for discovery, negotiation, delivery, and payment of data trades. The key challenge moving forward will be the complete elimination of brokers for the coordinating these functions to realize a full decentralized marketplace.

12.2 The Road Ahead

To conclude this chapter and the book, we will have a look on what lies ahead for blockchain in CPS.

The first area that is likely to attract future exploration is lightweight consensus algorithms. Although various consensus algorithms have been designed that significantly reduce the blockchain overhead compared with POW, there is still demand for a lightweight consensus algorithm that provides a high level of security, incurs low bandwidth and computational overhead on the participating nodes, and reduces the delay in reaching agreement over the state of the blockchain. The existing proposals in this space have made significant progress in this direction. The future is likely to see further development of these approaches and more integration of such consensus mechanisms into mainstream blockchain platforms.

The second major area of investigation will be privacy. As we have seen across many parts of this book, blockchain introduces a level of anonymity for the users, yet malicious nodes can classify transactions and deanonymize users in blockchain. One interesting research direction is to study the privacy level of the blockchain for CPS and how to enhance the user anonymity in both public and private blockchains. The private chains offer limited privacy due to reliance on a TTP, which limits the trust decentralization benefits of blockchain. New privacy-preserving private chains need to be designed to enhance user anonymity. New approaches that shift the pareto front of the privacy/trust trade-off will be in high demand, with an increased focus on attribute-based verification and zero-knowledge proofs.

A related open area for future investigation is key management. To enhance their anonymity as they transact in value exchange networks, blockchain users employ different keys, which create the challenge of keeping track of and managing the keys for each user. It is critical to design a lightweight key management framework that enables the CPS devices with limited computational resources to manage their keys while reducing the reliance on TTP. How to ensure that key management strategies do not leak any privacy-sensitive information will also be important.

The third key direction on the road ahead is revisiting regulations. As outlined in Chapter 5, the immutability of blockchain makes it impossible to remove transactions and data stored in the blockchain. The data protection regulations such as GDPR demand the users to be able to experience the right for their data to be forgotten. Despite a number of efforts performed to enhance the compatibility of blockchain with GDPR, it is critical to design blockchains that can fully comply with GDPR and enable the users to have full control over their data. Application-specific requirements for new regulations for supporting more flexible, decentralized, transparent, and fair marketplaces will also be central to the adoption of blockchain in many cyberphysical application scenarios.

The interoperability of blockchain systems is likely to be in high demand as well. As the popularity of the blockchain increases, multiple blockchains have been developed offering various services (e.g., Ethereum and Hyperledger fabric) and range from public to consortium and private. Recall from Chapter 3 that chain management and hierarchical methods are widely employed to reduce the blockchain overheads. It is critical for the users to transfer assets from the child chains to the parent chains or from one blockchain framework to another, known as interoperability. Most existing interoperability solutions rely on TTP, which leads to centralization and privacy challenges. Thus, there is a need for decentralized interoperability models that allow information and value exchange across different blockchain networks.

A common theme to many of the future directions is to reduce reliance on a TTP. Most of the blockchain applications rely on TTPs to conduct particular actions; for example, in an energy trading, a TTP oversees the trade to ensure that both sides commit to their obligations. However, the inclusion of a TTP often defeats the purpose of using a blockchain, as it effectively centralizes power, communication, and trust at the TTP. It is therefore critical to remove or reduce the reliance on TTPs to achieve decentralized solutions.

Storage limitations are expected to be a bottleneck for blockchains in CPS contexts, given the sheer scale of IoT and CPS networks. Few works are conducted to reduce the size of the blockchain database while enabling the users to verify transactions and ensure the consistency of the blockchain. Creating such frameworks is an interesting future research direction that may also potentially reduce the storage cost involved in managing blockchain, making the technology more attractive for broad adoption in CPS applications.

The application of machine learning in blockchain is an interesting research direction that has recently attracted tremendous attention. The convergence of blockchain and machine learning can potentially address challenges in either of these technologies and ultimately improve the services offered to the users. Machine learning algorithms can be employed to address the fundamental challenges of blockchain (see Chapter 3). For example, a new consensus algorithm can involve a learning process instead of solving a puzzle. As shown in Chapter 7, machine learning algorithms can be used to study the user anonymity in blockchain and enhance the anonymity level of blockchain users. Machine learning algorithms rely on the data received from users. Blockchain can function as the underlying data transmission platform that can ensure security and integrity of the communicated data. As an example, in [1], we introduced the Internet of Mobile Energy, which employs blockchain as the underlying communication framework to collect and transfer the data of the users in a smart city. Machine learning algorithms are employed to identify the best time to charge/discharge an electric vehicle or the best route toward a particular direction. In [2], the authors discussed the potential interactions between blockchain and AI, where blockchain can provide more traceable and explainable machine learning, or machine learning can help manage security within a blockchain network.

A related future direction is to distribute the learning task using federated learning [3] and blockchain, leading to an autonomous network. Federated learning was introduced by Google and enables the low-resource devices, such as smart phones, to participate in the learning process. Blockchain can establish trust and provide auditability and transparency, which enables untrusted nodes to collectively solve a task.

Another interesting research direction for blockchain in CPS is the convergence of blockchain and edge computing. CPS devices produce a large volume of information, which makes data processing and transmission challenging. In recent years, edge computing received tremendous attention as a mean to push data processing closer to where the data is generated, that is, the edge of the network, for example, processing the data in the home gateway instead of sending it to a cloud. This potentially reduces the delay in receiving service and the associated packet overhead. Blockchain enables all participating nodes that have enough resources to participate in processing data at the edge of the network. This potentially increases the CPS autonomy, as most of the network tasks can be distributed among the participating nodes.

Finally, trust has been a prevalent theme throughout this book, both as a generic and cross-cutting requirement for blockchain in CPS contexts, and also as a specific design goal for CPS application sectors. While blockchain establishes consensus among untrusted participants, the trust in the data generated by the devices and trusting the identity of the users remain open challenges. To establish trust in the data, it is critical to consider the resources consumed and the associated delay. In identity trust, it is critical to protect the privacy of the users while enabling them to prove their identity. Investigations into the multiple facets of cyberphysical trust and how they integrate into blockchain systems are likely to attract significant attention into the future.

Clearly, there is still a lot of work to be done to realize the full potential of blockchain for CPS. These open challenges are outweighed by the immense potential benefits of blockchain for CPS that will deliver decentralization, traceability, trust, security, and anonymity at scale. CPS are naturally decentralized and increasing autonomous, which is highly compatible with blockchain's decentralized architecture and its increased automation capabilities with smart contracts. We therefore anticipate a bright future for blockchain in CPS systems and applications.

References

[1] Jurdak, R., A. Dorri, and M. Vilathgamuwa, "Internet of Mobile Energy: Towards Seamless Energy Trading Across the Transport and Energy Sectors," *arXiv preprint arXiv:2003.10085*, 2020.

[2] Salah, K., et al., "Blockchain for AI: Review and Open Research Challenges," *IEEE Access*, Vol. 7, 2019, pp. 10127–10149.

[3] "Federated Learning," https://ai.googleblog.com/2017/04/federated-learning-collaborative.html.

About the Authors

Ali Dorri is a research fellow at Queensland University of Technology (QUT), Brisbane, Australia. He received his Ph.D. from the UNSW, Sydney, Australia. He was also a postgraduate research student at CSIRO, Australia. His research has received over 2100 citations and his h-index is 14. His research interest includes blockchain, IoT, security and privacy, distributed energy management, and distributed systems. He has published over 25 peer-reviewed papers. Dr. Dorri served on the organizing committee of SDLT and BCCA and as technical program committee in 10 conferences including the International Conference on Blockchain and Cryptocurrencies (ICBC). Dr. Dorri has delivered five talks and tutorials, including a tutorial at ICBC.

Salil Kanhere is a professor in the School of Computer Science and Engineering at the University of New South Wales (UNSW) Sydney. His research interests are in the IoT, CPS, pervasive computing, blockchain, cybersecurity, and applied machine learning. His research spans problems across the entire spectrum of applications, architectures, algorithms, and technologies. He has published over 250 peer-reviewed articles and has delivered over 30 tutorials and keynote talks on these topics. His research has received over 9,300 citations and his h-index is 47. He has received seven Best Paper Awards. Dr. Kanhere is a conjoint researcher at CSIRO's Data61 and is associated with the Cybersecurity Cooperative Research Centre. He has held visiting appointments at the Institute of Infocomm Research Singapore, Technical University Darmstadt Germany, Graz University of Technology Austria, and University of Zurich Switzerland. He is the editor-in-chief of *Ad Hoc Networks* and serves on the editorial board of *IEEE Transactions of Network and Service Management*, *Pervasive and Mobile Computing*, and *Computer Communication*. He serves on the Steering Committee of the IEEE LCN and has extensive experience in organizing top conferences in the discipline including IEEE PerCom, IEEE/ACM CPS-IoT Week, IEEE ICBC, IEEE WoWMoM, and ACM MSWiM. Dr. Kanhere is a Humboldt Research Fellow and an ACM Distinguished Speaker. He is a Senior Member of the IEEE and ACM. He received an M.S. and a Ph.D. in electrical engineering from Drexel University in Philadelphia, Pennsylvania, and a B.E. in electrical engineering from Veermata Jijabai Technological Institute, Mumbai, India.

Raja Jurdak is a professor of distributed systems and the chair in Applied Data Sciences, as well as the director of the Trusted Networks Lab at Queensland University of Technology. His main research interest is dynamic network modeling. Networks and network science provide powerful representation of physical and logical relationships to gain insights into a broad range of systems, from communication and CPS and IoT to individual and entity relationship networks and interactions to guide value-creating decisions. He is particularly interested in delivering end-to-end trust in data and systems involved in distributed networks, using tools such as blockchain and distributed ledger technology, understanding network dynamics driven by mobility to forecast diffusion processes, such as disease spread, and sustainable in situ intelligence in networks.

From 2008 to 2019, Dr. Jurdak was at CSIRO, where he established and led the Distributed Sensing Systems (DSS) Group to become one of the leading large-scale sensing group worldwide. He currently maintains a visiting scientist role with the DSS Group. He has a Ph.D. in information and computer science and an M.S. in computer networks and distributed computing from the University of California, Irvine, and a B.E. in computer and communications engineering from the American University of Beirut. He has published over 170 peer-reviewed articles and received three Best Paper awards. His h-index is 34 and his research has received over 6,200 citations. He authored a book on wireless ad hoc and sensor networks in 2006. He serves on the editorial board of four international journals, including *Ad Hoc Networks*, and on the organizing and technical program committees of top international conferences, including Percom, ICBC, DCOSS, WoWMoM, ICDCS, RTSS, and IPSN. He is a conjoint professor with the University of New South Wales, a visiting scientist at CSIRO Data61, and a senior member of the IEEE. He was a Eureka Prize finalist in 2019, received the QLD iAward merit award in 2016, and the in 2011 received the Endeavour Executive and CSIRO Medal for Environmental Achievement.

Index

Recent Titles in the Artech House Information Security and Privacy Series

Rolf Oppliger, Series Editor

Attribute-Based Access Control, Vincent C. Hu, David F. Ferraiolo, Ramaswamy Chandramouli, and D. Richard Kuhn

Biometrics in Identity Management: Concepts to Applications, Shimon K. Modi

Bitcoin and Blockchain Security, Ghassan Karame and Elli Androulaki

Blockchain for Cyberphysical Systems, Ali Dorri, Salil Kanhere, and Raja Jurdak

Bluetooth Security, Christian Gehrmann, Joakim Persson and Ben Smeets

Computer Forensics and Privacy, Michael A. Caloyannides

Computer and Intrusion Forensics, George Mohay, et al.

Defense and Detection Strategies against Internet Worms, Jose Nazario

Demystifying the IPsec Puzzle, Sheila Frankel

Developing Secure Distributed Systems with CORBA, Ulrich Lang and Rudolf Schreiner

Electric Payment Systems for E-Commerce, Second Edition, Donal O'Mahony, Michael Peirce, and Hitesh Tewari

End-to-End Encrypted Messaging, Rolf Oppliger

Evaluating Agile Software Development: Methods for Your Organization, Alan S. Koch

Fuzzing for Software Security Testing and Quality Assurance, Second Edition, Ari Takanen, Jared DeMott, Charlie Miller, and Atte Kettunen

Homeland Security Threats, Countermeasures, and Privacy Issues, Giorgio Franceschetti and Marina Grossi, editors

Identity Management: Concepts, Technologies, and Systems, Elisa Bertino and Kenji Takahashi

Implementing Electronic Card Payment Systems, Cristian Radu

Implementing the ISO/IEC 27001 ISMS Standard, Second Edition, Edward Humphreys

Implementing Security for ATM Networks, Thomas Tarman and Edward Witzke

Information Hiding Techniques for Steganography and Digital Watermarking, Stefan Katzenbeisser and Fabien A. P. Petitcolas, editors

Internet and Intranet Security, Second Edition, Rolf Oppliger

Introduction to Identity-Based Encryption, Luther Martin

Java Card for E-Payment Applications, Vesna Hassler, Martin Manninger, Mikail Gordeev, and Christoph Müller

Medical Device Cybersecurity for Engineers and Manufacturers, Axel Wirth, Christopher Gates, and Jason Smith

Multicast and Group Security, Thomas Hardjono and Lakshminath R. Dondeti

Non-repudiation in Electronic Commerce, Jianying Zhou

Outsourcing Information Security, C. Warren Axelrod

Privacy Protection and Computer Forensics, Second Edition, Michael A. Caloyannides

Role-Based Access Control, Second Edition, David F. Ferraiolo, D. Richard Kuhn, and Ramaswamy Chandramouli

Secure Messaging on the Internet, Rolf Oppliger

Secure Messaging with PGP and S/MIME, Rolf Oppliger

Security Fundamentals for E-Commerce, Vesna Hassler

Security Technologies for the World Wide Web, Second Edition, Rolf Oppliger

SSL and TLS: Theory and Practice, Second Edition, Rolf Oppliger

Techniques and Applications of Digital Watermarking and Content Protection, Michael Arnold, Martin Schmucker, and Stephen D. Wolthusen

User's Guide to Cryptography and Standards, Alexander W. Dent and Chris J. Mitchell

For further information on these and other Artech House titles, including previously considered out-of-print books now available through our In-Print-Forever® (IPF®) program, contact:

Artech House
685 Canton Street
Norwood, MA 02062
Phone: 781-769-9750
Fax: 781-769-6334
e-mail: artech@artechhouse.com

Artech House
16 Sussex Street
London SW1V 4RW UK
Phone: +44 (0)20 7596-8750
Fax: +44 (0)20 7630-0166
e-mail: artech-uk@artechhouse.com

Find us on the World Wide Web at: www.artechhouse.com